THE INDIVIDUAL'S
GUIDE TO GRANTS

THE INDIVIDUAL'S GUIDE TO GRANTS

Judith B. Margolin

PLENUM PRESS • NEW YORK AND LONDON

Library of Congress Cataloging in Publication Data

Margolin, Judith B.
 The individual's guide to grants.

 Bibliography: p.
 Includes index.
 1. Research grants — United States — Handbooks, manuals, etc. I. Title.
 AZ188.U5M37 1983   001.4 4   83-2252
 ISBN 0-306-41309-4

The author and the publisher gratefully acknowledge the permission of the copyright owners to reprint the quotations that introduce the various chapters, in particular, the quotation on page 63, from *The Seven Laws of Money,* by Michael Phillips, Copyright © 1974, Random House, New York 10022, and the quotation on page 157, from *What Color Is Your Parachute?* by Richard M. Bolles, Copyright © 1978, Ten Speed Press, Berkeley, California 94707.

Also, the list of "manipulative verbs" on page 57 is reprinted, with permission, *from The Universal Traveler: A Soft-Systems Guide to Creativity, Problem-Solving, and the Process of Reaching Goals,* by Don Koberg and Jim Bagnall, Copyright © 1981, William Kaufmann, Inc., Los Altos, California 94022. The "time saving strategies" on page 69 are reprinted from *Do It Now,* by William J. Knaus, Copyright © 1979, Prentice-Hall, Inc., Englewood Cliffs, New Jersey 07632.

First Printing — April 1983
Second Printing — September 1983

© 1983 Judith B. Margolin

Plenum Press is a Division of
Plenum Publishing Corporation
233 Spring Street, New York, N.Y. 10013

Printed in the United States of America

To the memory of my father
Edwin L. Goldberg
If only he had lived longer
— or I had written faster —
he might have seen it all in print

PREFACE

This book is a work of conscience. It is the product of a long-standing feeling of obligation on my part to write something useful for a special group of people to which you probably belong—individuals who seek grants.

In my years as Director of the New York library of The Foundation Center,* each and every day I encountered numbers of individuals looking for grant money. Although I tried to be as supportive as possible, in the face of the particular problems shared by this group of library users, my own reaction was one of relative helplessness. Simply stated, most of the fund-raising guides, printed directories, and computer files purportedly created to serve the fund-raising public are of little or no use to *individuals* who seek funding on their own. These resources are directed

---

*The Foundation Center is the independent, nonprofit organization established by foundations to provide information for the grant-seeking public.

toward the nonprofit, tax-exempt agency, which is the most common recipient of foundation, corporate, and government largess. They are not designed to respond to the special requirements of the individual grant seeker.

In the applicant eligibility index, the *Catalog of Federal Domestic Assistance* defines individuals as "homeowners, students, farmers, artists, scientists, consumers, small-business persons, minors, refugees, aliens, veterans, senior citizens, low-income persons, health and educational professionals, builders, contractors, developers, handicapped persons, the physically afflicted." In short, practically everyone qualifies.

The term *individual* as I use it here also has several meanings. Of course, each one of us is a unique individual with something special to offer a potential funder. In a sense, every grant awarded is given to individuals, since it is individuals who will carry out the grant projects. I also use the term individual according to the more technical meaning—a person with no formal affiliation to an institution, who still seeks grants. The individual for whom this book is primarily intended has little or no experience with the grant-seeking process. He is unaware of grants for which he may be eligible or whether, indeed, he is eligible for any at all. Presently, he is either unaffiliated with a nonprofit organization, or, if he is affiliated with one, does not know how to make the best use of his relationship with a sponsor. He is not sure where to begin his quest for funding nor how to write a successful proposal. He does know that he has a good idea and he would like a grant either to begin or to continue working on that idea.

One of the purposes of this book is to demonstrate that it is possible for an individual to maintain his uniqueness while at the same time hooking up with mentors, umbrella organizations, sponsors, and other institutions for the purpose of seeking funds. As I will indicate, the individual grant seeker's association with sponsoring organizations may be on a continuum, so that he is able to retain his desired level of independence while still qualifying for funding. Once the individual builds a reputation through the receipt of awards, honors, and small grants, he may wish to "trade up" to a larger grant and then become more formally affiliated with an organization.

Hence, this book, although providing as many concrete recommendations as possible for the unaffiliated grant seeker, will also prove useful to representatives of tax-exempt organizations seeking funding for their agencies. Many of the facts and recommendations set forth relate to organizations as well as to individuals. Details are given on the creation and construction of a full proposal. Even if the grant seeker has to submit only a short letter of intent, it is helpful to understand the components of a successful proposal and reassuring to know that one can go back and select from these elements if he has to write a longer proposal at a later date.

The grant seeker is plagued by a series of questions to which he has few answers: How do I begin? What types of projects stand the best chance of being funded, and is mine among them? What do I need to know? Once I've identified a potential funder, what do I do next? This book is a guide to the reader in finding those answers for himself.

For most people, time is a precious resource. Invest that resource wisely. Allow adequate time to identify a likely source of grant money and to elicit as much information as possible about your prospective funding source. Do not waste your time sending out a blanket mailing to every funder whose name has ever been mentioned to you. Don't toil for weeks preparing a near-perfect proposal and then casually send it out to a carelessly compiled list of prospects. Do your homework carefully. Select your potential funders wisely. A properly worded proposal will be much easier to assemble after you have selected the right funding source. Indeed, if you've done your homework correctly, your proposal should almost write itself. Because you have selected your funding sources with such care, it follows logically that presenting your ideas to them will be much easier.

I hope this book will be useful to you in your endeavor to raise funds. Good luck!

Judith B. Margolin

Old Greenwich
July 20, 1982

ACKNOWLEDGMENTS

I would like to thank my husband, Harold Margolin. He suggested—no, insisted—that I write this book. He believed in it long before I really did. Most of all, he provided the financial and emotional support necessary.

I also thank my children, Adam and Jessica, for getting along perfectly well without me. I apologize to them for my too-frequent absences (sometimes even when I was physically present) and for occasional bouts of short-temper. I hope when they are old enough to read the book they will think it was all worthwhile.

I am grateful to Helen Weisner, Shirlie Schwarz, and Lauren Grieb for the excellent substitute mothering they provided. Without them, there would have been no time to write.

To my editor, Linda Regan, I would like to express my most sincere gratitude for the twin traits of empathy and tact, which she has developed to the highest degree.

CONTENTS

" . . . all that dollars can do . . . is to find (or help train) a competent and devoted individual and give him time and tools and a place to work."
—F. EMERSON ANDREWS
Philanthropy in the United States

Chapter 1 / INTRODUCTION

YOU *CAN* GET A GRANT

It was estimated several years ago that one out of five people living in the United States at the time was receiving some form of grant support.[1] Contrary to the commonly accepted belief that *foundations* do not make grants to individuals, the fact is that they do; but many of these grants are disguised as gifts to institutions. *Corporations,* too, support individuals, not only by means of scholarship programs for their own employees but also through the provision of a wide variety of services to others via a mechanism whereby grants are treated as business expenses. As for the *federal government,* a glance at the *Catalog of Federal Domestic Assis-*

tance will indicate the vastness of possibilities for the individual grant seeker. *State governments,* too, support individuals with various programs, many of which originated from federal dollars. Even wealthy *individual donors* sometimes fund other unrelated individuals, either by means of outright gifts or by bequest. With such a wide range of funding sources and so many possible choices, almost any significant grant idea should qualify for some sort of funding.

In essence, all grants are actually awarded to individuals even if only indirectly; funders sometimes simply delegate the responsibility for supervision to sponsoring institutions. In the words of the late F. Emerson Andrews, who during his lifetime was often called Mr. Philanthropy: "After all, we are all recipients of philanthropic aid, if we go to college, use a library, or attend an endowed museum."[2]

These facts coupled with my own research and experience in the foundation field suggest that you as an individual *can* get a grant if you have a good idea and if you introduce that idea in an effective manner to an appropriate funder. This book is intended to help the individual do exactly that by demystifying the process of applying for grants. Very specific and pragmatic advice is provided to increase your chances of being funded. Sources of information on grants to individuals are examined and terminology is explained. Emphasis is placed where my own particular philosophy dictates it should be: on finding the right funder for you. As I have stated frequently in my work as a librarian, seminar leader, and consultant to those seeking funds, the most beautifully worded proposal based on the most ingenious idea will not produce a penny in grant money if it is sent to an inappropriate funder.

In the past, it was assumed that the individual was too involved in his own work or research to be sophisticated in the areas of selecting funders, writing proposals, and applying for grants. Thus, he was defeated almost before he had begun to search for additional funding. This lack of expertise is a luxury the individual grant seeker can ill afford. His survival in a competitive marketplace depends on his ability to discover and to attract funders. In many fields of endeavor today, he must be an informed, knowledgeable grant seeker simply in order to continue working.

INADEQUACY OF EXISTING RESOURCES

Foundations

Until quite recently there was no comprehensive listing of foundation grants either currently available or previously awarded to individuals.* Traditionally, this has been an area that the compilers of Foundation Center publications avoided even though it was acknowledged that "information on foundation funding sources for individuals is one of the least documented areas in the field of private philanthropy."[3] The reasons for the long-standing hesitancy to tackle this problem are difficult to fathom. They derived either from the assumption that direct grants to individuals represented too tiny a percentage of total foundation philanthropy to be significant or that they involved issues that were too sticky in terms of the foundations' own fears of government scrutiny of their activities. Yet, the record of such grants is *very* significant to the individual grant seeker. No matter how reluctant the foundations are to share information on this subject, the grant seeker needs to know what funds actually are available to him.

Most of the studies on foundation support of individuals are quite old. According to research conducted by the Peterson Commission in 1968, only 3% of foundations at that time were awarding grants directly to individuals.[4] Of these, 62% were for scholarships or fellowships, 23% for research, 7% for creative endeavors, 7% for the needy, and less than 1% each for awards/prizes and travel/study. Only 17% of the foundations queried by the Peterson Commission acknowledged that they had *ever* made grants directly to individuals. They tended to be among the smaller foundations, representing only about 2% of total grant expenditures.

It is fundamentally irrelevant to the individual grant seeker what percentage of all foundation giving his funder represents. He must work

---

*The publication of *Foundation Grants to Individuals* has contributed greatly to filling in this gap.

with the pool of resources available, no matter how limited. He needs more documentation on the grants that actually *are* awarded rather than a listing of carefully compiled statistics to prove that such grants are insignificant, for even a small percentage of the total grants dollar adds up to significant support to the right grant seeker.

CORPORATIONS

If the resources on foundation grants to individuals appear inadequate, the studies on corporate performance in this regard are even sparser. In fact, documentation of corporate giving on the whole is quite poor, except in the areas of the arts and education. In a survey sponsored by the Business Committee for the Arts and conducted by the Conference Board, $405,284 or about 6% of the total of $5,115,117 went directly to the individual artist. Of this figure, $385,216 was handled as business expense, $4,200 as corporate foundation grants, and $15,868 as outright corporate gifts.[5] It is extremely difficult to get reliable figures on corporate contributions since so many that actually fit the definition of grants are reported as business expenses. Most lists of corporate contributions, furthermore, exclude gifts in kind, especially donated time. As with foundation grants, there is a need for more comprehensive studies of corporate support of individuals. We know for sure only that there is a general trend to overall growth through the years.[6]

GOVERNMENT AGENCIES

I was unable to find any statistics or comprehensive listings of federal, state, or local government grants to individuals as opposed to funds earmarked for institutions or municipalities. This is in spite of the fact that the federal government has been a major provider of individual grants over the years, especially via the "government foundations," the National Foundation for the Arts and Humanities and the National Science Foundation.

INDIVIDUAL DONORS

To the best of my knowledge, gifts to individuals from unrelated individual donors are not documented anywhere. No matter how charitable their intent, they are not tax deductible and, therefore, not deemed appropriate for public scrutiny.

So much for the primary source material on grants to individuals. In preparation for this book I read scores of available guides to proposal writing and grant application. Almost without exception, these, too, were geared toward the organization rather than the individual. The authors of these guides assume that the reader is affiliated with a tax-exempt agency, has a prior knowledge of fund raising, a familiarity with the jargon of grantsmanship (itself an overworked and little understood word), and a wide range of institutional backup support, such as accounting and legal services, at his disposal.

My own experience indicates that this simply is not the case. From what I have observed, these guides are being read by an audience that differs markedly from the one for which they were intended. They are being used by individuals who are uninitiated in the work of applying for grants and confused by all the assumptions made that do not apply to the individual grant applicant.

The inadequacy of existing information sources is a sorry state of affairs that presents a serious problem to the grant seeker. In fact, to the more frustrated among us, it appears that a silent conspiracy has been in effect to keep us from getting precisely the information needed before we begin our quest for a grant, namely: Are funds really available? Is it worth the effort?

My answer to both of these questions is "yes!" Foundations, corporations, government agencies, and even individual donors *do* support individuals in the form of outright grants, awards, prizes, loans, and the provision of services. You will find examples of such funding throughout this book. Most of us know someone who presently works on a project funded by grant money. In my day-to-day research, I often encountered individuals who would say, "I've already had a Fulbright and a Guggenheim. Now I'm trying to get a Danforth," or words to that effect. To quote the

President of The Foundation Center: "Special requirements for private foundation grants to individuals set forth in the Tax Reform Act of 1969 discouraged giving in this area. However, a surprisingly varied group of foundation programs of this kind do exist, although with special limitations."[7]

EXPLANATION OF TERMS USED

Throughout this discussion there are certain terms that will be used with some frequency and, therefore, require elucidation at the outset. Other definitions are provided as needed throughout the book.

Individual has been somewhat delineated already. To reiterate: I use the term according to its several meanings. In a sense we are all individuals, since we are all unique. Every grant awarded benefits individuals, since it is individuals who will carry out the grant projects. In its more technical sense, the word refers to the unaffiliated grant seeker applying for funding on his own. However, I also use it to refer to the individual who is either unaware of affiliations he may have or uninformed about how to make use of his relationship with a sponsor in order to qualify for funding. The assumption is made that the individual who reads this book wants to retain some degree of independence while still getting a grant.

By *grant,* I mean any gift of money from a funder to an individual recipient. A grant is not the product of an isolated act with little precedent or carry-over. Rather, the word brings with it the implication of an ongoing *process,* including seeking out the funder, applying to the grant maker, and using the funds in a responsible manner once received. There are, however, many services and vehicles of support other than cash grants that funders provide. Thus, the discussion also refers to awards, prizes, loans, and nonfinancial support—anything that will benefit the individual and help him to get started. There are various types of grants that are discussed in later chapters. This is not a "scholarship" guide. There are many of those already on library shelves. Much of the material on getting into gear, proposal writing, and grant application, nonetheless, will be useful to students seeking financial aid.

By *grantsmanship,* another word you will see mentioned, I refer to the body of knowledge and series of activities whereby the individual becomes a credible applicant and a responsible recipient of a grant. Grantsmanship is neither an art nor a science but an accumulation of methodology through much hard work and the process of trial and error.

Funder is used here to refer to any organization or other type of donor that makes grants to individual recipients. Something to be remembered about funders is that the large institutional ones are made up of human beings with faults and virtues. Funders covered in this book are foundations, corporations, government agencies, wealthy individuals, and miscellaneous associations.

The term *idea* is used in its very broadest sense to describe the concept that you want to demonstrate or establish by means of a grant. It may be a project or a series of ideas; it may be an experiment; it may be a performance, a product, or a gallery showing. Whatever your idea, the assumption is that in some way it is original and that you are qualified to carry it out. Of course, some ideas are more likely to be funded than others. Getting a good idea is discussed in further detail later on.

One last point on word usage: while writing this book I was disturbed by the question of gender. After much internal debate, I decided to adopt the traditional masculine form. In my readings, I came across the following paragraph that expresses my sentiments exactly:

> Finally, like many of today's writers (and editors), I find myself stymied by the third-person-pronoun problem: Should I say *he* or *she?* Should I torture the sentence into an awkward shape in order to avoid the problem? Should I repeat a noun endlessly so that personal pronouns are unnecessary? No groundbreaker by nature, I have fallen back on the convention of using the masculine pronoun, with apologies to those it may offend, and with the hope that some brave and imaginative soul, be he a she or a he, will soon solve the problem for all of us.[8]

WHAT THIS BOOK CONTAINS

The purpose of this book is not to bemoan the inadequacy of existing resources or to make a plea for more comprehensive study of grants to individuals, rather, it is to suggest how to exploit the resources available to you. As I hope you will discover, they are many. Throughout the discussion, the emphasis will be on your adoption of a new view of yourself not as a petitioner but as a wise investment for the funder.

First, you will determine whether or not you need institutional affiliation. Second, you will confront the most painful and difficult evaluative task—judging whether or not your ideas truly merit funding. Third, you will reorient yourself in the best possible shape for seeking a grant. Following this will be an explanation of the idiosyncratic patterns of the various types of funders and specific recommendations on finding the right funder for you. You will then consider the ingredients of a proposal and reflect upon what makes some more successful than others. Finally, you will contemplate what your responsibilities will be if you are funded and how to proceed if you are not. All of this information is presented in a manner geared especially to you—the individual grant seeker.

NOTES

[1] *Annual register of grant support* (10th ed.). Chicago: Marquis Academic Media, 1976.

[2] Andrews, F. E. *Philanthropic giving.* New York: Russell Sage Foundation, 1950. Pp. 250–251.

[3] *Foundation grants to individuals.* New York: The Foundation Center, 1977. P. i.

[4] Report and Recommendations of the Commission on Foundations and Private Philanthropy. *Foundations, private giving and public policy.* Chicago: Univ. of Chicago Press, 1970.

[5] Chagy, G. (Ed.). *Business in the arts '70.* New York: Eriksson, 1969.

[6] Fremont-Smith, M. R. *Philanthropy and the business corporation.* New York: Russell Sage Foundation, 1972.

[7] *The foundation directory* (6th ed.). New York: The Foundation Center, 1977. P. xxiv.

[8] Balkin, R. *A writer's guide to book publishing.* New York: Hawthorn Books, 1977. P. xiv.

"... human beings in general, and most of the time, need each other—even desperately. An individual cannot be described or understood, nor can he live his life isolated from the society of which he is a part."

—ROGER J. WILLIAMS
You Are Extraordinary

Chapter 2 / NO MAN IS AN ISLAND

WHO SHOULD READ THIS CHAPTER

On initial consideration, you might be tempted to skip right over this chapter about affiliation. After all, hasn't it already been stated that thousands of grants are awarded each year by various types of funders directly to unaffiliated individual applicants? Isn't the main premise of this book that it is possible to apply for, receive, and utilize grant money all on your own? The answer to these questions is yes, yes, *almost* unequivocally yes. And yet ... There are individual applicants who surely could benefit from some form of institutional affiliation. A number of grant ideas truly are enhanced by organizational sponsorship. Others more or less require it. To cite a few examples:

- If you are an inventor working on an idea for a gadget that will improve the quality of everyday life, but you do not require the use of high technology equipment or laboratories, you can probably apply to a government agency or foundation on your own.

• If you are a mathematician whose grant idea relies on complex computer programming and the assistance of technical personnel, you probably do require affiliation with an academic institution or research institute. Funders will want to know that you have access to the appropriate technology before they will award you a grant. They have to be sure that you possess the wherewithal to do what you propose to do.

• If you are a sculptor whose idea involves creating a statue to improve the appearance of a local park, and you already have your own studio, materials, and tools, you most likely can apply to a local corporation or family foundation directly on your own.

• If you are a poet who proposes to give readings at various elementary schools "to turn kids on" to poetry, you might find it helpful to have the local school board sponsor you. This will reassure potential funders that there is actually interest in your idea at the community level, that you will have official cooperation, and that your schedule will be coordinated in a systematic way.

• If your idea involves writing a work of fiction, and the grant you require is merely for living expenses, then you can apply directly to funders that make such grants for works in progress. Should you have a publisher, a letter of endorsement from your editor helps. If your manuscript is complete, you can apply to one of the associations that give awards under such categories as "best short work by a new author."

• If your idea is to conduct an oral history project by interviewing on tape those who lived through a natural catastrophe, you probably should establish a formal connection with a nearby historical society. Funders will want to know that you are working under the auspices of a respectable agency that will supervise your procedures, seeing that you are in touch with appropriate subjects for your interviews.

• If you propose to work on a new medical theory that certain enzymes combat a particular disease, sponsorship by one of the major medical research bodies is essential. There are thousands of controversial theories about various diseases, many coming from apparently competent scientists. Grants decision makers ordinarily do not themselves have the technical expertise to oversee your idea from a scientific standpoint. They need to know that some qualified, prestigious body will attend to this for them.

• If you propose to travel outside the country either to gain or to share information in your field of specialization, you definitely must apply under the auspices of a fellowship program or foreign-exchange agency. In this instance, the sponsor serves to guarantee the funder that you are not just taking a paid vacation and also promises to supervise you during your stay.

• If you are a social worker, psychologist, or other type of counselor and you propose to operate a telephone hotline, a youth shelter, an alcoholism program, or other counseling service, of course, you can apply on your own to a foundation or corporation for initial funding as well as equipment rental. However, if you find a church, community center, or local agency like the YMCA to sponsor you and perhaps to donate some space, it will lend authenticity to your idea, which in turn will make it more attractive to funders.

This list could be continued ad infinitum. Determining whether or not you need a sponsor is up to you. The decision rests on whether you as an applicant can make it on your own, whether you *want* to go it alone, whether your idea lends itself to development by just one individual, and whether backup services are required. There are as many types of sponsors as there are ideas for grants. As the previous examples indicate there are varying levels and degrees of affiliation with any given sponsor. The choices concerning what kind you need and how structured or loose the relationship should be are also up to you. The individual grant seeker can and should create a sponsoring arrangement that benefits him and his grant idea, *while not losing sight of the motivation of the funder.*

WHY MANY GRANTS CALL FOR INSTITUTIONAL AFFILIATION

The arguments in favor of affiliation are no doubt already familiar to you. It is said that fewer than one out of thirty foundations and "very few"[1] government agencies are willing to consider proposals from individuals without institutional affiliation. Corporations are reluctant to make grants to such individuals unless they are related to the corporation in some way and/or a direct line may be drawn with the best interests of a particular business. Even in cases where grants *are* made directly to individuals, there is often a nonprofit organization, institution, or university "hovering in the background somewhere."[1] Before you become discouraged by these facts, keep in mind that although at least 90% of all grants are awarded to organizations, many of these grants—and we don't know how many—include hidden sponsoring arrangements whereby new groups and individuals qualify for funding under an established sponsor's name. If you were to catch sight of these donations in a grants list, you would have no way of knowing, for example, that the $5,000 grant to the Junior League was actually for one member to complete and publish a booklet on local child-care services. Also there are countless grants and awards given directly to individuals without formal institutional affiliation, which have never been formally tabulated. Thus, the statistics regarding direct assistance to individuals are somewhat misleading.

The main reason that funders—foundations, corporations, and government agencies—hesitate to give directly to individuals is simple: it costs too much! Reportedly, it takes as much administrative time and work to award a $5,000 grant to one individual as a $500,000 grant to a university. The obvious administrative strategy to keep down the cost is to give a lump sum to one outstanding group (a sponsor) whose function it is to "pass [grant monies] on down the line."[2]

A second reason that funders prefer affiliated applicants is that sponsors serve as a kind of buffer to dilute the funder's responsibility in case something goes wrong. Such responsibility is of particular concern to foundation administrators, whose programs of grants for individuals have been the subject of much Congressional scrutiny over the years. Most foundation grants are awarded to tax-exempt organizations, which in turn select individual recipients to participate in various programs. Foundation managers thereby defuse their own responsibility for the selection and supervision of individual recipients and are not as answerable in a direct way to the Internal Revenue Service (IRS), even in the case in which grantees abscond or misuse the funds. Such cases are rare, indeed, but they have been known to happen.

In the field of scientific research, most grants are made to support the work of individuals. Yet almost without exception, these individuals have formal institutional affiliation of one sort or another. Hence, a third reason for the affiliation requirement: Sponsoring institutions provide a wide range of facilities, equipment, backup, and administrative services, unavailable to the individual lacking such affiliation. In today's technologically advanced society, many new ideas for scientific and other grant projects require relatively complicated procedures, too complex for one individual to conduct in his own surroundings without access to specialized facilities.

For the above reasons, many funders (but certainly not all) do have policies against direct grants to unaffiliated individuals. Individual applicants must apply *under the auspices* of a university, or other tax-exempt organization, or in some cases, under a special committee—that is to say, a sponsor. If the grant is approved, the funder's check is made out to the sponsoring organization, which in turn administers the project, doling out funds to the individual as needed.

For the unaffiliated individual, these remarks are not intended as discouragement. If you decide to apply on your own, it is important that you understand the traditional reasons that funders have preferred applicants with institutional affiliation. They are based on real constraints upon the funder and are not just arbitrary.

The lesson to be learned is that individuals who seek grants should be familiar with the restrictions on such funds. If you can, by all means apply on your own, but concentrate on funders that have programs free from such restrictions or find ways to work around them. If this is not possible, a small dose of institutional affiliation certainly cannot hurt you. As you will see later in this chapter, sponsors are not hard to find. In fact, you probably already are affiliated with some organization or agency that could perform this function for you. With a little ingenuity you can make it easier for a funder to give you a grant without compromising yourself or your ideas. To quote William H. Whyte, Jr.: "A sensible awareness of the rules of the game can be a condition of individualism, not merely a constraint upon it."[3]

THE AFFILIATION CONTINUUM

One of the catchwords of the grant seeking world is *affiliation*. It is most probable that when you began to consider the possibility of seeking a grant, one of the first things you were told was that you must be "affiliated" in order to receive a grant. This statement is unnecessarily intimidating to the grant seeker. As already indicated, thousands of grants are awarded each year directly to individuals without specific institutional affiliation. Just to list a few:

A tape collector has received grants of $10,000 from the National Endowment for the Arts and $15,000 from the New York State Council on the Arts to amass a recording-tape library of various machine sounds.

The John D. MacArthur Foundation, which was formed upon the death of the insurance wizard, awards up to fifty annual "no-strings-attached" prizes of $25,000 to $45,000 for five years to fellows nominated by a squad of talent scouts. The concept behind these grants is to set brilliant individuals free to work on their own ideas on a variety of subjects.

The U.S. Department of Energy conducts a program whereby inventors and small businessmen receive grants of varying amounts to develop and market their own energy-saving ideas.

At a summer Connecticut music festival, there were two composers whose works were on the program: one composer was acknowledged to be a Fulbright scholar and the recipient of a Guggenheim fellowship and awards from the Koussevitzky Music Foundation and Boston Musica Viva; the other, also a Fulbright fellow, was the recipient of a National Endowment for the Arts Composers Grant and the Martha Baird Rockefeller Recording Award.

The Institute of Current World Affairs awards one or two fellowships of about $10,000 a year to individuals (usually in their twenties or thirties) to observe and study at firsthand foreign cultures or issues of contemporary significance. Grant money covers living and travel expenses for two to four years.

The Fund for Investigative Journalism awards about fifty grants each year, ranging from $100 to $2,000, to journalists pursuing specific investigations of a controversial or a worthwhile nature.

This list could be continued almost indefinitely. The grants listed were chosen more or less at random not because they are indicative of particular trends in grant making but rather because none of the individuals who have received or will receive these grants are "affiliated" in the sense that their grant money was contingent upon their having formal sponsors. Yet all of the recipients have many affiliations with various groups, organizations, and institutions in their daily lives. In this sense, no one is truly "unaffiliated."

As the title of this chapter suggests, none of us works alone on a desert island. Defining affiliation in its broadest context as "belongingness," we all belong somewhere. Determining which of our present contacts or those available to us can best serve as sponsors for grant projects requiring such sponsorship is one feature of successful grantsmanship for the individual grant seeker.

Affiliation may be viewed as a continuum ranging from working almost totally in isolation to becoming an employee of a nonprofit institution in order to seek funding for your idea (see Table I). Identifying the exact degree of affiliation necessary to your success is an essential preliminary step in the development of your grant proposal.

There are numerous soft edges and overlaps on the affiliation continuum. Your job is to find where you most comfortably fit in. The following discussion will help you determine where on the continuum your idea belongs and the minimum degree of affiliation required to help you accomplish your goal—to get a grant.

WORK ON YOUR OWN

At one end of the affiliation continuum is the individual whose idea can be developed without either institutional backing or much recourse to the skills and expertise of others. Those in the fine arts—poets and fiction writers, in particular—may fall under this rubric. Also in this category are inventors with simple ideas that can be accomplished within a short time period at one location. Another group is applicants for various awards and/or prizes, e.g., architects competing for cash prizes for best design of a local firehouse; scholars applying for travel funds to present a paper at an international conference; and medical researchers, as well as those in other fields, applying for awards in recognition of past achievement or demonstrated excellence in given subject areas.

Nevertheless, even in these cases of direct grants to unaffiliated indi-

TABLE I
THE AFFILIATION CONTINUUM FOR THE INDIVIDUAL GRANT
SEEKER

Level of affiliation	Examples
Work on your own	Inventors Sculptors Poets Photographers
↓	
Form a consortium with other individuals; develop affinity groups; form your own tax-exempt organization	Self-help groups Local ecology groups Sports clubs Chamber orchestras
↓	
Find a temporary "in name only" sponsor or umbrella group, to serve as a fiscal agent or conduit	Arts councils Historical societies Church groups Film societies United funds
↓	
Make use of a current affiliation to serve as your formal sponsor, or find a new sponsor	Professional societies Trade associations Clubs Unions Alumni groups
↓	
Become an employee of a nonprofit institution	Universities Libraries Hospitals Museums

viduals, some type of influential connection is usually beneficial. It lends credibility to the applicant if the funder knows he was referred, recommended, or even nominated by an academic or other institution or by a mentor, teacher, supervisor, or a prominent member in his field or community. In other words, even if you decide to go it alone, your valuable personal and professional connections still can be helpful. Do not overlook them.

Although the pleasures and rewards of working on one's own are many, the individual applying for such grants may be plagued by doubts and fears that serve as obstacles to the development of his idea into a full-fledged grant proposal. First and foremost among these obstacles is a sense of isolation. The grant seeker applying on his own may feel especially isolated because of the lack of an existing support system, compared to institutionalized philanthropy with its conferences and journals, development specialists and computer files, official and unofficial workshops/seminars, and other support groups.

To overcome feelings of isolation, the grant seeker should attempt to carve out his own personal support system. It is hardly ever valid to say, "I don't know anyone." We all have "connections" of some sort. Sit down and take the time to explore yours. Whom do you know who can help you? Whom do you know who knows someone who can help? Talk to that person or persons.

Other obstacles familiar to most grant seekers are jealousy of those already funded, competition with other applicants, and fear that someone will steal your ideas. Combat these destructive attitudes by adopting a new stance of cooperation. Again, communication is desirable. Talk to people. Work on the assumption that there *is* enough to go around. Remember that fears and self-doubts are normal reactions to the grant-seeking process. Grant seeking involves the taking of risks, which is almost always a painful process. However, most of your fears will lessen once you take the plunge.

Many grant seekers, be they individuals or representatives of large organizations, experience one final obstacle—shame of needing money. This is an anachronistic carry-over from the long-standing American tradition of rugged individualism. It is merely a question of attitude and can be easily changed. Bear in mind that we all need some form of support from our fellow man. Society, perhaps most of all, requires the support of grant seekers who serve as the suppliers of new ideas and the facilitators to turn their new ideas into valuable products or services.

FORM YOUR OWN ORGANIZATION

Closely related to applying for grants on your own—and another option available to the grant seeker along the affiliation continuum—is forming your own (usually nonprofit) organization. Such organizations may be loosely or formally structured depending upon the degree of affiliation required by potential funders. They range from affinity groups of individual grant seekers who band together to provide one another simple moral support to formal tax-exempt organizations incorporated for the purpose of seeking grants for specific projects.

If your idea involves the solicitation of funds from the general public as well as from funders (e.g., a matching gift or similar situation whereby funders provide only part of the project money), if it involves a construction program, renovation, or the large-scale purchase or rental of equipment (e.g., the transformation of a vacant inner city lot into a park, playground, or community garden), if it requires complex long-term procedures calling for several staff people in addition to yourself (e.g., an archeological dig, a free mental health clinic, a dramatic performance), then you probably should consider forming your own organization.

Forming your own organization may take time (from three months to one year), money (from $100 to several thousand dollars in legal, accounting, and/or registration fees), and may be unappealing to the individual grant seeker who shuns officious formalities. Obviously, not all ideas lend themselves to the establishment of nonprofit organizations. For those that do, nonetheless, this is another possibility for the grant seeker. It serves as an alternative to seeking institutional affiliation and sponsorship.

The concept of forming an organization is one of establishing a coalition to create funding opportunities where none may have previously existed. You become, in effect, your own sponsor. Stated simply, if funders require institutional affiliation, you'll give them an affiliation, in the form of your own nonprofit tax-exempt institution.

If the organization you form qualifies for a certificate of tax exemption from the IRS, it is worth its weight in gold in terms of grant dollars, because a certificate of exemption is required by many corporations and foundations prior to consideration of your proposal. For information on securing tax exemption, see IRS publication 557, *How to Apply for and Retain Exempt Status for your Organization* (Washington, D.C.: U.S. Government Printing Office).

A useful dictum to follow in forming your own organization is always to get competent professional advice. Secure free advice where available from those who have already done it and information from agencies like the Small Business Administration—an excellent resource on starting your own business and related topics. You may not have thought of your grant idea as a small business, but to quote A. Vernon Weaver, head of this government agency: "Anyone who makes a product *is* a small business." Getting paid professional help as required from lawyers, accountants, bankers, and others provides a kind of insurance and usually proves economical in the long run, because it helps prevent serious mistakes.

To become familiar with the nonprofit incorporation procedures of your own state, write or call your Secretary of State for formal requirements, application forms, and fees. Be aware in advance that the process will take time and will vary enormously in terms of complexity from state to state.

Even though relatively few individual grant projects lend themselves to this type of formal incorporation, it is one option available to you if you wish neither to work entirely on your own nor to seek an institutional sponsor. If you do decide to take this route, a useful book is Fred A. Whitaker's *How to Form Your Own Non-Profit Corporation in One Day* (Oakland, Calif.: Minority Management Institute, 1977). Although this publication will get you started in the right direction, it is nearly impossible to complete the process without some recourse to professional legal advice.

UMBRELLA GROUPS

Forming your own organization need not imply the establishment of a large, formal institution. It may mean simple coalitions, consortia, or affinity groups that have no formal, separate legal status. Such group formations and alignments may be quasi-independent from existing community organizations or technical assistance agencies or may function as arms or chapters of such agencies. Any nonprofit organization is eligible to serve as an "umbrella group" for the purpose of applying for grants. By umbrella group, I am referring to an intermediate agency, usually nonprofit, that receives and disburses funds to individuals. Churches, schools, community organizations, self-help groups, arts councils, and even local clubs, if they are nonprofit, may serve as umbrella groups for grant applicants.

Examples of the kinds of individual projects that might benefit from this type of less formal arrangement are seasonal events, such as festivals, craft demonstrations, art shows, workshops, training seminars, or poetry readings, all of which are executed within a relatively short period of time. As an individual, or a group of individuals, you really don't need to form an organization to implement such ideas. You simply require some existing group to cover for you (as an umbrella) for the duration of the grant project. Such a short-term affiliation may take the form of a project under the auspices of an umbrella group or of a simple offshoot from it.

SPONSORS

It is fair to say that since most grants are awarded to institutions rather than directly to individuals, one of the major obstacles to an individual applicant's receiving funding is not knowing how to go about finding an institutional sponsor to apply for his grant.

How can you tell when you need a sponsor? Here are some specific cases where sponsorship of an individual grant applicant is indicated:

- If your idea for a grant requires access to a laboratory, studio, classroom, accounting, or secretarial or computer services, you may find you need an institutional sponsor.
- Should your idea be controversial or involve some element of risk, funders might shy away from it unless you find a sponsor to endorse the idea and lend you credibility.
- If the idea you have in mind will have substantial projected impact on your community, then you probably need a reputable

community agency as your sponsor. It will serve to reassure funders about the changes you may bring about in your own backyard.

In each of these examples, the determination of whether or not affiliation with a sponsor would be beneficial rests on the nature of the grant idea combined with the requirements of the funder.

Sponsors have been defined as "groups, agencies, and institutions with established performance in your field, a record of solid visibility, accomplishment, and thus high-quality credibility."[1(p.151)] To put it more succinctly, sponsors are third-party groups that collaborate with you to meet funders' requirements. You provide the idea and the manpower; the sponsor provides the track record and credibility. Together you form a kind of symbiotic relationship.

The following types of organizations may serve as sponsors for individual grant seekers. Many have done so in the past: schools, colleges, universities, research institutes, educational associations, professional societies, state and local art associations, historical societies, museums, hospitals, health agencies, nonprofit performing groups, sports clubs, scientific societies, social and recreational clubs, fraternal organizations, community foundations, unions, labor, agricultural, and horticultural organizations, veterans' groups, civic leagues, chambers of commerce, and churches and religious groups of all types and sizes.

This list is not comprehensive. I merely mention some of the types of sponsoring organizations here to give you an idea of the broad range of groups that is available. It probably has become evident to you by now that you are *already* affiliated with several groups that might serve as sponsors for your grant project. In the rare instance in which this is not the case, the list is provided as a jumping-off point to help get you started in determining what type of organization would best serve you in this capacity. Of course, the most important criterion for an effective sponsor is that this organization *enhance,* not detract from, the development and funding potential of your grant idea.

What to Look For in a Sponsor

Whether you decide to use one of your current affiliations or to seek out one that you have never before contacted, the selection of the right sponsor is crucial because funders carefully examine institutional factors before they award you a grant. They will question whether the local setting and conditions are appropriate, whether available facilities like laboratories, storage, and support services are adequate, and especially whether your grant project seems to have strong community support. Since one advantage of organizational sponsorship is that it provides you with credibility, it is important that you select a sponsor with a good reputation and a rapport with its own funders. The only way to ascertain this is to make discreet inquiries among staff or board members as well as others who are knowledgeable about the various sponsors in your field.

Also, examine the services the sponsor presently provides, if any. Do they have a satisfied clientele, community support, and the cooperation of the local political structure? Determine which segments of the population are most directly affected by the sponsor's activities. Speak with representatives of these groups. Make contact with local community leaders to find out if a potential sponsor is well thought of and if its services are effectively targeted and appreciated by those who use them.

At first this kind of detective work may make you feel distrustful or even underhanded. However, remember, there is a good deal at stake here. In casting your lot with a sponsor for a period of time, you are closely allying your own name and reputation with that of the organization. Sometimes this connection is difficult to sever and may last in the minds of others (funding executives in particular) after the actual sponsoring arrangement has broken off. So select your sponsor carefully. You don't want your own reputation and credibility jeopardized for future grants.

As a sponsored individual, you may not be privy to many of the goings-on within the institution. It is, therefore, important that you choose a sponsor whose top management is sympathetic with the implementation of your ideas. Ideally, the leadership of the sponsoring organization should view your grant project as a beneficial extension of the organization's own services.

Limited Role of Sponsors

You may get the sponsor in some instances to act in a limited capacity as a fiscal agent. Under this type of sponsoring arrangement, the organization performs a channeling or conduiting function. It assumes expenditure responsibility for the grant money, removing the burden somewhat from the funder. Take note, however, that because of stringent IRS regulations regarding grantor responsibility, many funders are reluctant to give grants to conduits, and many sponsoring organizations hesitate to function as conduits.

The most common conduits are church groups, fund-raising arms of schools, colleges, and hospitals, and various community organizations. The role of the sponsor in this type of arrangement is generally limited to financial matters only, and this type of fiscal arrangement is sometimes also referred to as *in name only* sponsorship. The sponsor accepts the grant, since the check is made out to the organization and allocates it to the intended individual recipient as needed. The sponsor may or may not keep track of your expenses as you proceed. As a grant seeker for various types of projects, you may prefer this kind of limited sponsoring arrangement, if you can find it, since it guarantees you some degree of independence.

Advantages and Disadvantages of Affiliation for Grant Seeker and Sponsor

Whether you are seeking out a new sponsor or trying to convince some group with which you presently are affiliated to serve in this capacity, you should stress the advantages of this arrangement when you approach potential sponsors. Since you may have to sell your grant idea to a funder later on in much the same way that you sell yourself to a sponsor, consider this as a sort of preapplication exercise.

The advantages to the organization of sponsoring you will vary depending upon the nature of your grant idea and the type of sponsor. The best way to begin the process of selling your idea is to sit down and make a list of what advantages you can foresee for potential sponsors. Here are some suggestions of possible benefits to sponsors to help get you started:

1. By sponsoring your grant project, the organization in question may attract new funders to its own programs thereby making it possible to bring to the attention of potential funders other ser-

vices in addition to your grant project the sponsoring organization may provide.

2. The grant money you bring in may help to spread the sponsor's own overhead charges and defray some indirect costs.

3. If your project is successful, the sponsoring organization shares the honors or benefits from your ideas.

4. Serving as a sponsor for your grant project may enable the organization to expand its own services or programs with a minimal financial outlay from its own annual budget. (The desk, phone, laboratory, and so on are already there.)

5. In some instances a sponsor may pre-empt the competition that you might present to them were you to go elsewhere with your ideas.

Some of the advantages to be gained by you, the individual grant seeker, when affiliating with a sponsor have already been touched upon: increased availability of otherwise restricted grant funds, enhanced credibility and prestige as an applicant, and access to a wide range of institutional facilities, equipment, and support services. Other possible bonuses are better working conditions and experienced professional assistance in such areas as publicity and fiscal management. Whereas some projects require your spending considerable time administering the funds, affiliation with a large institutional sponsor can turn some of the managerial duties over to the sponsor. An additional incentive is the medical, life, and disability insurance that some associations provide at relatively low group rates to those they sponsor.

There are, of course, disadvantages to a sponsoring arrangement, many of which are endemic to the concept of "collective work" in a highly technological society. Many individual grant seekers fear that sharing their labors will inhibit the creativity and originality of their proposed project, and this fear is somewhat justifiable. For along with affiliation comes the possible loss of autonomy and the resulting sense of obligation. Such indebtedness may be either explicit or implicit in a sponsoring arrangement.

According to an old saying, "there is no such thing as a free lunch." This is particularly true of sponsoring arrangements. If a group becomes your sponsor, it's a pretty safe bet that the leaders of that organization will expect *something* in return, be it servility on your part, reflected prestige, enhanced public image, or some kind of a finished product (e.g., a report, a flow chart, a booklet, or a training module).

How to Find a Sponsor

As indicated earlier, start with your current affiliations. Make a list of clubs, professional associations, educational institutions, and work-related groups with which you are presently or were formerly affiliated. Ask yourself if any of these groups would be useful either as sponsors or as potential sources of information on who else might sponsor you. Speak with your friends and contacts at the various organizations and with their leaders. The president of the club, the dean of the college, and the pastor of the church are all useful sources of information on how to proceed. They may even be flattered that you seek their advice. When you reach an impasse or receive a negative response, however, always ask for a referral: "If you can't help me, whom would you recommend?" Once someone cannot help you directly, you may find yourself with the psychological advantage of the other person's guilt. Why not make use of it?

If your present affiliations provide no fresh leads to possible sponsors, the next step is to look in your own backyard. Approach the traditional nonprofit organizations in your own locale. Depending on the situation, you might call on local hospitals, churches, public television and radio stations, civic groups, YM and YWCAs, historical societies, and even such groups as the United Way. To make optimal use of each personal contact, be sure to ask for further referrals.

Along with this local search for a sponsor, you should also conduct a detailed search in your subject area by thoroughly investigating your own field. Refer to current grants lists detailed in Chapter 6 to see which organizations in your own area of specialization are presently receiving foundation, government, and/or corporate support. Make contact with the leaders of these organizations. You may be surprised to find how cooperative they will be. Ask them how they went about securing funding, what advice they can offer you, and with whom you should speak. Perhaps one of these recipient organizations might even serve as your sponsor. Since these groups already have positive relationships with funders, their leaders may welcome the opportunity of taking on additional projects for future grants.

Next, speak with your colleagues, others in your field, and, especially, with the experts. Again you may be surprised that a well-known individual will take the time to speak with you on the phone or answer your letter with some valuable insights.

An excellent resource for the grant seeker in his quest to locate potential sponsors is the *Encyclopedia of Associations* (Detroit, Mich.: Gale Research Company). This reference guide details over 15,000 national, trade, professional, and other associations, many of which could qualify as sponsors for individual grant seekers. A number of these associations themselves award small grants, scholarships, and fellowships, run contests, and bestow medals, certificates, and trophies, all of potential benefit to the individual. Use the *Encyclopedia* as a source for referrals in your search for a sponsor. Call or write associations in your field to determine who's doing what, who has been funded, and who has a good reputation. Once you've established your identity as an individual with similar interests, you may find the leaders and staff of such associations more than willing to chat with you.

Use the geographic index of this national reference guide also to identify local chapters of large organizations that might serve as your sponsor. Trade associations—organizations, clubs, institutes, and societies (e.g., groups of people who do the same type of work you do and are interested in the same subject areas)—are invaluable resources when it comes to information both on funding possibilities and on sponsorship opportunities.

Structuring Your Relationship with a Sponsor

Once you have identified a potential sponsor, the next step is formal affiliation for the purpose of seeking and receiving a grant. How loose or structured your relationship with the sponsor will be depends upon the individual situation and the customary practices of the sponsor. Some will treat you more or less as an employee, requiring you to work on the premises, some as a consultant, some as an affiliate, and some will leave you pretty much alone. Ask about such administrative ties before you make any promises so that you can select a sponsor whose practices in this respect will most nearly coincide with your needs.

Since you may require professional guidance from your sponsor in administering the grant as well as seeking funds, it is always best to avoid surprises. Ask in advance. Secure in writing, if possible, what is expected of both you and the sponsor and who supplies what services in exchange for what, right down to the last pencil.

To avoid being gobbled up by your own sponsor, consider drafting a letter of agreement at the time you decide to affiliate. Better yet, have an attorney do it. Such a letter functions as a contract if you both sign it. It specifies the responsibilities and benefits of both you and the sponsoring organization as a direct result of your joint collaboration. Ordinarily, it sets a time limit on your affiliation. Remember, sponsors will want you to be quite specific about your plans once the grant money runs out.

For a sample letter of agreement between an individual and a sponsoring organization, see Smith and Skjei.[pp.59–60]

BECOME AN EMPLOYEE

At the opposite end of the affiliation continuum from working on your own is one final stopping off point—becoming an employee. This is an alternative rarely considered by the individual grant seeker. Yet it is a perfectly viable option. In certain specific instances, it is really the only option, since some types of grants are awarded solely to institutions, e.g., grants for building or renovation, salaries, general operating expenses, or the purchase of equipment. In these cases and others, the more informal sponsorship arrangement simply is not enough to qualify the individual for funding. If the potential funders you've identified make grants only to organizations, or if the idea you have in mind is one for which only an institution may apply for a grant, you may decide to join an institution.

Becoming an employee of a nonprofit organization need not be as intimidating as it initially appears. On the one hand you *do* stand to lose some degree of control over the direction of your ideas. On the other hand, you stand to gain a great deal in terms of steady pay for a period of time, medical insurance, pension plans, and other fringe benefits. The greatest fringe benefit of all, of course, is access to all the resources at the disposal of the institution, not the least of which is grant money. When seeking out a potential institutional employer, follow many of the same criteria that you used in seeking a sponsor. The type of work environment you should look for is one that stimulates the creative side of the worker and permits competent individuals to pursue their own areas of inquiry, not just the work of teams. Talk to regular employees and other individuals who are conducting grant projects under the auspices of this organization. Ask about working conditions. How much of the day-to-day work time is involved in attending meetings, conferences, administrative details, and other forms of team interplay?

The ideal situation would be selecting an organization that already has grant money and excellent ongoing relationships with several funders and convincing that organization to hire you as a paid employee to work on your idea. This is not as far beyond the realm of possibility as you might think. Others have performed similar feats in the past.

What is more likely, however, is that you will find some nonprofit agency active in your field of endeavor or in a related field, that has not had outside funding before but whose leaders are anxious to seek grants in your field of specialization. Various types of community organizations and arts groups in particular fall under this category. New groups starting up and those branching out might be particularly interested in seeking funding. Occasionally, even larger organizations like libraries, museums, hospitals, or educational or research institutions might agree to become employers of those seeking grants.

Before you can get your own grant project off the ground, however, as an employee you may have to spend some time working with the leadership of the institution to identify funders, write proposals, and build up an ongoing grants-seeking program (all of which will be discussed in subsequent chapters). You become, in effect, their temporary in-house grantsperson. In my experience, the situation just described is not an uncommon occurrence. Although you stand to lose some time for working on the development of your own grant idea, the experience and contacts gained may prove invaluable to you and your career.

Keep in mind that full-time employment need not be forever. Such an affiliation need be no longer than your brief relationship with a sponsor. The employer will probably want you to guarantee that you will stay with the organization for the duration of your own project or until funds run out. However, when your tenure is over, you can go on to other projects. In the meantime, you can view the work within the institution as a sort of "paid graduate school,"[4] where you learn new skills and prepare for work on your own while collecting a salary to boot. If you decide to become an employee, an excellent resource is Richard Nelson Bolles' *What Color Is Your Parachute?* (Berkeley, Calif.: Ten Speed Press, 1982).

CONCLUSION: INDIVIDUALISM AND THE GRANT SEEKER

To paraphrase a statement John Steinbeck once made: Great ideas come only from individuals working on their own. American society as we know it today evolves from a firm belief in rugged individualism whereby each individual pulls himself up by his own bootstraps. According to prevalent but perhaps unfair stereotypes, inventors are natural recluses; scientists jealously guard secret formulas from one another; artists require uninterrupted solitude in order to work; performers are temperamental and unable to get along with their compatriots; grant seekers are noted pariahs, eternally in competition with one another, and unwelcomed by funders because of their numbers and incessant demands.

Yet, in apparent contradiction to the tradition of rugged individualism, a popular doctrine among funders today is that only second-rate thinkers work in private.[5] Grants decision makers favor projects involving established organizations and ideas that affect large numbers of people. They are impressed by coalitions among grant seekers, especially when they exhibit an awareness of what others in their field are doing and a willingness to touch base and cooperate whenever possible with other grant seekers and funders as well. Like a judge setting bail, a grants decision maker employs one major criterion when evaluating a proposal—the strength of the individual grant seeker's roots in the community in which he works. Hence, the unaffiliated grant seeker must stress those aspects of his grant project that enable him to cooperate, instruct, or help others in order to make himself and his idea more attractive to funders.

The endeavor to resolve the apparent conflict between the American tradition of individualism and the current emphasis on collective work among grant seekers is intrinsic to this discussion. It *is* possible to work "on your own" on your idea for a grant while simultaneously seeking the support of others. Such balance is difficult but not impossible to achieve.

No One Works Alone

William H. Whyte, Jr., in *The Organization Man* said that the ideal affiliation limits total involvement in any organization to the minimum necessary for performance of essential functions.[3(p.233)] The grant seeker must strike the right degree of compromise between his freedom to work on his own and the demands placed on him by other individuals and institutions. This compromise may be exemplified in part by the clash beween the purity of his own ideas versus the possible meddling influence of the funder's own requirements, which eventually may diverge from his own. When the concept of affiliation is initially introduced to the grant seeker, it can be quite frightening. He may immediately conjure up visions of being swallowed up by vast organizational structures and of someone else stealing or taking credit for his ideas. Yet, as an individual grant seeker, it is important that he learn not to view himself as a victim of current trends in grant making. Affiliation with a sponsor need not stifle his work but should serve as an adjunct to his attempt to seek funding.

David Reisman and other sociologists have asserted that an individual's *real work*—the field into which he would like to throw all his emotional and creative energies based on his own gifts and character—cannot coincide in this day and age with what he gets paid for doing.[6] My response to this assertion is—why not? I believe that the individual *can* stem the tide of collective work by viewing institutional affiliation in the proper circumstances as a positive, rather than a negative, component of grantsmanship.

NOTES

[1]Smith, C. W., & Skjei, E. W. *Getting grants.* New York: Harper & Row, 1980.

[2]Dollard, C. *Memorandum on grants to individuals.* New York: Carnegie Corp., 1938. P. 8.

[3]Whyte, W. H., Jr. *The organization man.* New York: Simon & Schuster, 1956.

[4]Weaver, P. *You, Inc.* New York: Doubleday, 1973. P. 8.

[5]Belcher, J. C., & Jacobsen, J. M. *A process for development of ideas.* Washington, D.C.: News Notes and Deadlines, 1976. P. 8.

[6]Reisman, D. *The lonely crowd.* New Haven, Conn.: Yale Univ. Press, 1950. P. 325.

"But if you do not express your original ideas, if you do not listen to your own being, you will have betrayed yourself. Also you will have betrayed your community in failing to make your contribution to the whole."

—ROLLO MAY
The Courage to Create

"Even the best of ideas is doomed if time and money are not available to push it to fruition."
—JAMES L. ADAMS
Conceptual Blockbusting

Chapter 3 / IN THE BEGINNING IS AN IDEA

WHAT MAKES A GOOD IDEA?

The entire grant-seeking process in a sense may be viewed as finding the means to facilitate good ideas. Now that we have examined the possibilities of affiliation and degrees of independence that a grant seeker may strive for, let us look at the step-by-step process of getting a grant. This chapter will help you learn to evaluate your own ideas as potential vehicles for grants.

Consider the following examples of discoveries and inventions all of which developed from good ideas:

- Although it cannot be verified, legend has it that millions of years ago one of our ancestors, more ape than man, rubbed two sticks together, thus setting off a spark. The spark gave him an idea. Fire was born, and the course of history was irrevocably changed.

• The wheel was the idea of another early anonymous ancestor living in Southwest Asia, who may have been attempting to transport stones more quickly from one location to another.

• Around 1800 B.C., some free-thinking individual was to change the course of civilization by coming up with the idea of the flush toilet. It was then installed in the royal palace at Knossos, Crete. However, this device inexplicably disappeared for quite a long time. Patented in London in 1755, it did not gain acceptance in America until the 1870s.

• In 1886, John S. Pemberton, a maker of patent medicines, had an idea for a new drink which he intended as a tonic for nerves and a remedy for hangovers.[1] He introduced it to his cronies at a local soda fountain in Atlanta, Georgia. They liked it, suggested certain changes, and gave it its name—Coca-Cola.

• Albert Einstein first advanced his theory of relativity in 1905 when he was only 26. His idea that matter and energy are interchangeable, not necessarily distinct, is the basis for the development of the atomic age and resulted in new concepts of time, space, motion, and gravitation.

• In the early 1950s, Dr. Jonas Salk had an idea which, from a medical standpoint, was highly unconventional. It occurred to Dr. Salk that several injections of a killed virus might confer immunity. Thus, he developed the polio vaccine, nearly eradicating a disease that was killing and maiming children all over the world.

What do these ideas have in common that made them worthy of pursuit? What makes these ideas "good?" Are they inherently valuable, or is it the way the originators of these ideas put them into effect? Do ideas simply come to their "owners," or do such individuals have to work hard to produce them? Can you truly "own" an idea? Is it necessary, or even possible, to protect your ideas from others? This chapter will address these and other questions.

SOME DEFINITIONS

IDEAS

For the purpose of this discussion, the word *idea* is very broadly defined. When I speak of "your idea for a grant" I refer to something you may be already working on, or more likely, dreaming about. Almost everyone has had an idea, at one time or another, that he felt might be truly valuable, if only he had the money to execute it. Many good ideas have never gone beyond the thinking stage because of lack of funds. Others were best left as mere daydreams. Still others have been brought to fruition through grant money. Some have been successful. Some have not.

It would be helpful if there existed a simple rule of thumb to tell in advance whether your idea would make a good grant project. Unfortunately, no such rule exists, at least to my knowledge.

Evaluating ideas is difficult for both you and the grants decision maker. There is nothing scientific about this part of the grantsmanship process. The decision to go ahead and seek funding as well as the decision to fund a given project are truly subjective ones. To judge your idea as fairly as possible, however, try not to fall in love with it. Too much emotional involvement can warp your judgment. Though it is impossible to be truly objective about your own ideas, keep in mind that every good idea should carry with it some means of built-in evaluation. Ask yourself if your idea is truly capable of success. Consider at the outset: How will you and the funder know if you've succeeded? Have you chosen the best possible solution? Get in the habit of being your own devil's advocate. Acting as your own worst critic, you may well prevent rejection by potential funders. Furthermore, by questioning your own ideas, methods, and solutions in advance, you may end up improving upon them.

Ideas have been defined as "products of individual human minds." The word *individual* is important here. Organizations and institutions cannot have ideas. Having an idea is a highly personal act that an individual does on his own. Without ideas, organizations would cease to function. In this sense, ideas have been viewed as the "heartbeats of institutions."[2]

SOLUTIONS TO PROBLEMS

Another view of ideas defines them in a more concrete way as problem solvers. According to this definition, every new invention may be viewed as filling a need and every successful idea as providing for something previously lacking. Ideas, then, are formal blueprints for solving problems. Problems in this context may be viewed as questions to which you address yourself. Your idea for a grant would be a proposed solution to the problem you have identified.

RELATED TERMS

Your written proposal (discussed in detail in Chapter 7) may be viewed as a mature, refined idea. A project might be thought of as the execution of your proposal, presumably once you have received grant funds.

Other terms closely allied to the process of having new ideas, if not synonymous with the word *idea* itself, are *design, experiment, innovation, novelty, original concept, unique procedure,* and, especially, *invention, creation,* and *implementation.* Invention often implies the design of a gadget or the development of a theory that makes one's work easier, faster, or better. Creation may be defined as the combination of seemingly disparate elements to form a functioning, useful whole. Implementation refers to the crucial second stage during which the individual thinker (in this case, the grant seeker) takes an idea, selects an appropriate application for it, and then carries it out. For a grant seeker to accomplish this requires funding.

WHAT KINDS OF IDEAS ATTRACT FUNDERS?

To attract funders with your idea, it is best to keep in mind an important dictum: "evolutionize; don't revolutionize."[3(p.30)] Successful grant ideas must have some relation and continuity with past ideas while still providing solutions to current problems. The solutions need not be new or even innovative. In fact, funders historically have preferred low-risk projects with solutions involving the same or similar methods that were used by previous grant recipients. What should be fresh about the solutions you propose is the style in which you apply them to the problem or the manner in which you intend to implement them.

The best way to determine whether or not your idea would truly solve an existing problem, fill a gap, or correct a shortcoming in society is informally to survey your field, something that inventors refer to as "checking out the market." Find out what others in your field or in related areas are working on. There is no need to discover new ideas for a dying field. Don't propose something that is too advanced for public acceptance. By the same token, don't come up with ideas that are too sophisticated for serious consideration by grants decision makers not particularly savvy in your area of specialization.

An informal premarket test of your idea is an essential step in the beginning stages of proposal preparation; it will save you much time and grief later on. Unless reliable sources and good evidence indicate that a need really exists for the service or product you want to provide, or unless you consider the work itself to be its own reward, you can save yourself future aggravation by realizing that your idea is interesting but "not feasible at the present time."

Once you've established that there truly is a market for your idea, you should next consider whether a particular funder has a need for it. Consider the funder's motivation. Ask: How will my idea help fulfill the goals of this institution? When you identify the answer to this question, stress that aspect as you work on developing your idea for any written or oral proposals to potential funders. For different funders, you might want to stress different aspects of your idea.

Timeliness is of the essence when transforming ideas into fundable projects. In grant making as in other fields, certain types of ideas are in vogue at a given time. Some ideas simply are not suitable for funding because the timing isn't right. They should be shelved and, if you continue to believe they are worthy, bring them out again ten years from now or when conditions are more suitable. However, don't limit yourself only to areas already funded. Most funders are receptive to new ideas— if they are good ones—even if they fall outside stated limitations. There's always room for some innovation.

Must your idea be truly original or unique to attract funders? Keep in mind that new ideas are everyone's business. Many inventions seem obvious once someone else has already manufactured them. We've all had the experience, when confronted with something new, of saying to ourselves—now, why didn't I think of that? Similarly, many of us have seen our own idea become the basis for highly successful inventions, discoveries, and grant projects in the hands of others, whereas we, unfortunately, did not act on it when we should have. So don't torture yourself with doubts about whether or not your idea is truly "new." Your own idea takes flight quite naturally from the ideas of others when the timing is right. Indeed, there are noted instances of simultaneous discovery or invention. Witness the case of Charles Darwin and Alfred C. Wallace who conceived the idea of natural selection at approximately the same time, each independently outlining his views in separately published essays in 1858.

HOW THE PRESENT GRANTS SYSTEM MAY HAMPER THE GROWTH OF NEW IDEAS

The impact of new ideas as possible vehicles for grant monies depends just as much on the person judging them as on their intrinsic worth. Because of increasing costs of management and technology, the grants decision maker must choose carefully among ideas to decide which ones he will invest money in and who will carry them out. Moreover, his professional reputation is at stake when he gambles on a new idea. Yet, the funder's reluctance to waste money on a new untried idea may be superseded by the fear of losing money or prestige, or both. This occurs when a competitor backs an idea that turns out to be a success. Although funding executives communicate and work with one another, they are also in competition.

Some disgruntled grant seekers maintain that the grants system itself presents a major drawback to the development of new ideas, because funds are awarded by administrators who are rarely, if ever, "idea men" themselves. According to this view, grants decision makers are like loan officers at a bank. They tend to be an unadventurous lot who seek a nearly guaranteed outcome, and this tendency has a negative impact on the development and selection of new ideas for funding. To quote one disconsolate grant seeker: "The specter of the grant administrator is an unwholesome apparition during the critical but vulnerable period in which an individual assembles his ideas and formulates his plans."[4(p.296)]

To digress a little, there may, in point of fact, be a natural clash between the type of person who becomes a funding executive and the creative individual who seeks grants for his new ideas. This clash is exemplified by the difference between "right-handed thinking"—associated with law, order, reason, logic, mathematics—and "left-handed thinking"—associated with beauty, sensitivity, subjectivity, imagery, humor.[5(p.38)] Administrators of funding institutions generally may be classified as "right-handed thinkers"; many grant applicants tend to be "left-handed thinkers." Bridging the gap between the two divergent points of view necessitates ingenuity, compassion, and sometimes as much creativity as required to come up with a new idea for a grant in the first place.

Researchers have identified a cluster of characteristics commonly shared by creative individuals. Included among these are the childlike traits of fantasy, mental playfulness, daydreaming, and questioning the obvious. Also characteristic of creative individuals is a willingness to take risks, coupled with an almost missionary zeal. Whatever new idea he happens to be working on seems totally and utterly essential and consuming to the creative thinker. Such a strong inner conviction supersedes parental, peer, and even supervisory approval.

Creative individuals possess great fluency and flexibility of thought processes, well-developed senses of form, and powerful imaginations. They also seem to be freer of self-doubt than others. Able to go against the mainstream, they are not as afraid of appearing foolish. If one were to isolate a single common denominator typical of creative individuals, it is the ability to discover hidden similarities that connect previously disparate frames of reference.

There has unfortunately been a tendency among certain segments of our society (some grant administrators included) to view creativity as an expression of neurotic patterns. Rollo May and other theorists disagree with this view. They insist that talent is not a disease; creativity is not a neurosis. Rather, it is evidence of a quest for form amidst chaos.

Having new ideas for grants, therefore, requires courage, and creativity, but frequently carries along with it an irrational sense of guilt.

In the words of Degas: "A painter paints a picture with the same feeling as that with which a criminal commits a crime." Since new ideas almost always destroy something, be it established belief, earlier hypothesis, or merely the status quo, those who have them sometimes suffer from a superstitious fear that their creative acts will provoke the jealousy of the gods.

Along with the grants decision makers who hesitate to try out new ideas, the leaders of organizations that sponsor you or institutions that employ you may do their part to stifle creative experiment. Such sponsors and institutions are tied not only to the "financial apron strings" of corporations, government, and foundations but to their own special interests as well. Some grant seekers would maintain that the creative individual is less hampered by poverty or the poor opinions of others than by grant restrictions imposed by "the shadowy presence of enthusiastic sponsors peering over (his) shoulder while at work."[4(p.296)]

At the same time, those seeking grants on their own, without sponsors, may well confront a curious bias against them. Some funders view unaffiliated grant-seeking individuals with suspicion. They may ask: If you're so good how come you don't have a reputable organization behind you? or How is it that someone else hasn't already proposed your idea for a grant? Confronted with these attitudes, the grant seeker may begin to believe that whether he affiliates or not, he just can't win.

Although so many elements contrive to inhibit the expression of new ideas, each year many new projects—the brainchildren of various individuals—do get funded, for good ideas tend to build their own momentum. Trust them to thrust themselves into the limelight sooner or later. Your role is to serve as a kind of midwife at the birth of a full-blown grant project from a vaguely formulated idea.

Above all, try not to be discouraged. Rejection by a funder does not necessarily signify that your idea is "bad." Remember that there may be a natural antithesis between the kind of person who has new ideas and the kind of person who makes decisions whether or not to fund them. Suggestions for presenting yourself in a more favorable light to such decision makers are provided in later chapters.

HOW TO DEVELOP YOUR IDEA FOR A GRANT

When you begin the grantsmanship process, it ordinarily is assumed that you already have a good idea for which you are seeking funding. However, what if the idea you have in mind turns out to be impractical or otherwise unattractive to funders, and you want to improve upon it, revise it, or develop a new one? What if you would like to seek funding, but you don't have a specific idea for a grant, just some vague thoughts or a particular field in which you are working or would like to work?

The purpose of this section is to help you generate better ideas and to improve upon the ones you already have. Many individuals experience blocks to the creative idea-having process because of the emotional problems that accompany it. This section will examine such blocks and offer techniques for overcoming them.

HINTS FOR GETTING STARTED

There are a number of helpful hints to aid you in priming the pump to prepare yourself to produce a good idea. When you first begin to develop your idea for a grant, don't stray too far from your own field of expertise. New ideas for grant projects are most likely to derive from your profession, hobbies, or leisure-time activities. Insights often arise when you least expect them or when you are working on something seemingly unrelated. Keep a pocket notebook with you for jotting them down.

A useful stimulus for developing ideas is to work on a problem or question immediately before going to sleep. Several possible solutions frequently present themselves upon awakening. Keep another notebook or pad near your bed so you can immediately write down your ideas or else they may be easily forgotten.

There are other mechanisms you can employ to help you expand upon your idea. Most of all, don't try to rush the process. Remember that creativity is highest after a problem has been thoroughly studied and digested. Sometimes drawing at random on subconscious thoughts, associations, or memories produces new ideas. Also, do not discount the value of subliminal learning. It may appear as though you are going through a nonproductive phase, but you really may be picking up a great deal from your surroundings and the work of others. Open yourself up, become receptive, read, observe, and listen carefully until you spy a gap—a need for a new product or service or a problem related to your area of interest that needs solving. Then permit your new idea to emerge at its own pace.

CREATIVITY ENHANCEMENT TECHNIQUES

The theory behind creativity enhancement as expressed by Gilbert Kivenson in *The Art and Science of Inventing* is that "undisciplined but highly innovative thought processes can be systematically channelled to produce streams of problem-solving concepts."[6] In terms of grant seeking, this suggests that it is possible to harness your apparently amorphous creativity in order to generate specific ideas for grants that either solve problems or fill gaps. In order to perform this feat, you should adopt certain techniques or procedures for making idea-having or problem solving easier and more direct. You can develop and improve your creative tendencies through specific exercises, mental games, and plain hard work. Nearly everyone can learn the habit of developing his ideas by means of training and practice. What follows is a brief description of some of the techniques you can use to enhance your own creativity, help you have more and better ideas for grants, and develop, expand upon, and improve the ones you already have.

Relax the Mind

Learning to relax the mind is a key to facilitating better conceptualization. Letting go of some of the rigid controls of logical thought processes decreases the inhibiting effects of what has been called the *ego* and *superego* and permits one to employ some of the subconscious techniques mentioned earlier. In a sense, you are tricking your mind into coming up with a new idea. Many magicians, politicians, and other professional persuaders are adept at this technique.

Sometimes a certain "mental playing around"[7] or permitting your mind to wander and free associate with no commitment to a fixed way of doing things will lead to an entirely new approach for a problem at hand. It is important that you not follow the path of highest probability. According to Edward de Bono in *Lateral Thinking,* perversity or deliberately selecting the wrong path, often leads to a better point of view. Young children are familiar with the type of puzzles called *mazes.* In this type of puzzle, more often than not, in order to help the dog reach his bone, it is necessary to follow a seemingly dead-end trail or the one fraught with the most apparent obstacles and detours.

Harvest Chance

One of the basic tenets of lateral thinking is learning to "harvest chance." Obviously, no one can control chance. However, de Bono believes the individual can create an environment that may favor successful results. To make the best use of chance, he recommends that the individual adopt a "passive but watchful" attitude or readiness of mind.

Closely related to harvesting chance is being in the right place at the right time, or quite simply, exploiting your own good luck. Creative individuals sometimes appear to be luckier than others. Perhaps this has to do with the flexibility of their thought patterns. Just as they are able to harvest chance, they recognize and seize the opportunity when luck presents itself.

The British scientist Arthur Fleming was able to adapt a lucky break to his own and the world's advantage. He returned from vacation to discover that he had neglected to throw away some contaminated laboratory slides prior to his departure. As he glanced at the slides in his wastebasket, he noticed that the mold growing on them was killing off the bacteria he had so carefully cultivated. In this accidental way, penicillin, first called *mould juice,* was discovered.

Visualization

De Bono also stresses the importance of selecting the proper problem-solving language for generating ideas. The choice of the right kind of language is important since ideas tend to beget other ideas. He makes a plea for alternate languages to the verbal and mathematical ones with which we are most familiar. He recommends the language of the senses, in particular, what he refers to as *visualization*. Visualization, in this context, means the ability to *see* your idea (and its subsequent implementation) in the mind's eye. Closely related to visualization is the ability to draw pictures as visual expressions of your ideas, a talent de Bono feels everyone should develop. For example, Nikola Tesla, called by some the "Father of the 20th Century," is said to have developed his visualizing abilities to such an extent that he envisioned the complete model of the first full-fledged induction motor, a sketch of which he scrawled in the sand with a stick while going for a Sunday stroll in a Budapest park.[7]

Change Your Viewpoint

There are times when simply changing one's viewpoint in an arbitrary manner will result in the birth of a new idea. All that is necessary is a shift of emphasis. Concentrating on a different aspect of a problem may result in a sudden flash of insight that was previously inaccessible to you because your focus was elsewhere. Thomas Edison, for instance, was stymied in his effort to develop a telephone that could transmit the human voice without static. Quite arbitrarily, he forced his attention away from complex questions of transmissions to the purely mechanical one of the operation of the telephone button itself. He then tested every single chemical in his laboratory until he came up with the right combination to make a carbon button that worked, static free.

Mental Games

Changing your viewpoint involves role playing that might at first seem absurd. Don't be afraid to make use of whimsy and zaniness within yourself. For starters, you might imagine yourself to be an object, any object having to do with your idea. If you are capable of this type of imaginary leap, sometimes an entirely new adaptation of your idea presents itself. A further variation of this technique is to indulge in deliberate reversal: imagine the tail wagging the dog. The invention of the conveyer belt came from an individual who imagined a highway moving while vehicles remained stationary.[8]

Many of the techniques involved in creativity enhancement are like games. The practice of "mental doodling" is apropos here.[9] Beginning daydreams with "what if . . ." can be a useful means of coming up with new ideas for grants. Doing something the "wrong" way, just to see what may happen, more often than not brings into view a whole new perspective. The trick is to relax and enjoy it. Leonardo da Vinci was a mental doodler par excellence. Starting with premises like "What if man could fly? . . . float down to earth like a pod? . . . swim under water like a dolphin?" he developed ideas in the 15th century that served as the basis for the modern helicopter, parachute, and submarine.

List Making

Another technique almost unanimously favored by creativity specialists is the making of lists. Lists ensure good definition of a problem and some permanency of your ideas simply by the fact that they are written down. List making stimulates freedom of thought, because it postpones evaluation and judgment on your part, permitting your ideas to flow. Simple checklists are useful thinking aids as well. They help you to pinpoint what you may have overlooked and to expand on your present idea.

De Bono insists that having things too rigidly defined tends to stultify creativity, since categorizing ideas too soon is exclusionary and inhibits the birth of related new ideas. One trick to avoid premature labeling of your idea is to make a list of the separate attributes of a problem. This forces a breakdown of your idea into its component parts, and frequently new patterns emerge that were previously camouflaged.

To adopt the technique of list making, first concentrate on the problem or gap for which you are developing an idea for a grant. Make a conscious effort to relax your mind. Suspending all evaluative criteria, make a list of every thought coming into your mind related to your idea. Don't discard any thoughts. Write them all down, no matter how silly. Put away your list without reviewing it for two or three days. When you go back to it, you may find that your subconscious mind has done its work in between. The hook up of three or four seemingly unconnected thoughts might just result in a new approach. Seldom will full-fledged, coherent ideas pop up during the list-making process. They usually appear later while you are working on something entirely different.

BLOCKS AND OVERCOMING THEM

As a postscript to this discussion of techniques for improving your ability to generate good ideas, I would like to touch upon those elements within yourself and your environment which may block your creativity. So far, the major roadblock to coming up with grant ideas that we've mentioned is the individual's own fear of presenting his ideas to funders and being rejected. A second type of inhibiting force is exemplified by what James L. Adams in *Conceptual Blockbusting* calls *conceptual blocks* or mental walls that prevent problem solvers from correctly perceiving a problem or conceiving its solution.[5(p.11)]

At one time or another, you have probably experienced a vague sense of guilt about a particular task that never seems to get done. Either you have found something more important or, even worse, more enjoyable than the task at hand, or it has been too distasteful to start that day, so you have kept putting it off until the next tomorrow. Similarly, every single time you've begun work on a particular problem, you've discovered that you're stuck—stymied by one detail you can't seem to get around or one difficulty you can't surmount. You may have had an idea in mind, but you couldn't begin work on it until a particular circumstance in your life changed, e.g., you moved to a bigger house, your divorce became final, you switched jobs. All of these are illustrations of *blocks* to creativity, which greatly inhibit grant seekers.

I mention these blocks not to discourage you. On the contrary, once you recognize them, you will better be able to cope with them, work around them, or even harness them to your own advantage. Creative activity is a mental process subject to the influence of many complex psychological forces. Becoming aware of certain drives and subconscious influences helps you to overcome your emotional immobility. Isolating your own fears, anxieties, and hesitations helps disrupt those subconscious processes that distort or limit creativity. Identifying these blocks is the first step toward doing away with them:

Perceptual Blocks. Improper identification or delineation of the initial problem or gap you and your grant idea strive to solve. Perceptual blocks involve inadequate emphasis or focus, covering too large or too narrow an area, as well as the inability to see an issue from various perspectives.

Cultural Blocks. Imposed by the society in which you live, these blocks result from the resistance of most cultures to change. Based on biases and taboos, they hamper the individual within a given society.

Environmental Blocks. Imposed by the immediate surroundings in which you develop your ideas, these blocks derive from physical surroundings that are not conducive to creative thought, e.g., your work place is too hot, too cold, too confined, too open, or too cluttered with too many distractions. A poor psychological climate can be just as inhibiting (or even more so) as in the case of unappreciative supervisors, distrustful colleagues, lackluster competition, and unsupportive staff.

Emotional Blocks. Personal conflicts that arise within the individual's own psyche. These include the inability to tolerate ambiguity, the tendency to judge rather than to generate ideas, inadequate motivation, and a broad spectrum of self-doubts and fears.

Escape Mechanisms. Avoidance, procrastination, diversion—turning to something that is less demanding of your powers of concentration—and distraction—turning to another new idea before you implement the one at hand.

How to Overcome Blocks to Creativity

You will find that many of the techniques for enhancing creativity are also useful for overcoming blocks to new ideas. In addition, the following suggestions may be adopted by you if you are stuck or find yourself blocked in your quest to generate a new idea or solve an important problem that serves as the basis for a grant proposal:

- Write a "catastrophic expectations" report.[5(p.55)] Ask yourself—what is the worst possible outcome if everything goes wrong? You may be surprised to find that failure would not be as terrible as you originally supposed. Very often, people do not assess the probable consequences of their creative acts in a realistic way.

• The best cure for procrastination is a stern and rigid work schedule. Be as tough on yourself as a boss might be. Conduct yourself as though you were in a business situation that requires you to be in a certain place at a certain time and to behave in a professional manner. If interruptions persist, think of the work on your grant idea as a magnet that draws you back, just as soon as circumstances permit.[10]

• To combat avoidance, diversion, and distraction, impose more discipline on yourself. Establish an initial set of rules. Write them down if necessary. The more creative thinking you do, the more your unconscious gets the message that it is okay, and the more creative you will be. A momentum develops that stimulates the imagination to produce a continuous stream of new ideas. If you don't act now, you run the risk of becoming a "yearner instead of a doer."[11]

• Stay with your idea. Eventually it may enable you to produce a great invention, a beautiful work of art, a new scientific theory, or a solution to an important problem that in turn will serve as the basis for a successful grant proposal.

NOW THAT YOU'VE GOT A GOOD IDEA, WHAT SHOULD YOU DO WITH IT?

Once you've developed your idea for a grant to its fullest potential, there are three basic processes with which you should concern yourself: writing it down, testing it, and selling it to a funder.

WRITING IT DOWN

A process nearly as painful for the grant seeker as evaluating your own idea for a grant is writing it down. The sources of this discomfort are twofold. In the first place, you fear that in writing it down some essential aspect of your idea will be lost, that in defining it you are somehow limiting it. Judith Groh in *The Right to Create* expresses the fear this way:

> For whenever a thought is captured in words, a precise inflex-
> ible backbone penetrates the rich, inchoate blur of imagination's
> art. Hoisted on its convenient mast, the idea can be graded, exam-
> ined critically, and shown to the world. But in the process, part of
> the idea may be lost.[4(pp155-156)]

A second fear that makes communicating your idea so distasteful is one of self-exposure. Writing down your thoughts requires you to create some order among them and to project your idea outside yourself. Self-protection, shyness, mental confusion, and uncertainty combine to make you postpone this trial of strength.

Part of your fear, of course, is that others will laugh at you, finding your idea to be impractical, trivial, or even ridiculous. Once it is written down, you yourself may find your idea to be less consequential than you had hoped. It is much easier to keep your idea snug within your own mind as a possible source of a grant if only you had the time or the expertise to apply for one than it is to write it down in black and white, thus transforming it into a proposal for all to read. After all, an idea that exists only in your head cannot be rejected.

Despite your inhibitions about putting pen to paper, if you can't write down your idea in a sentence or two, it is still too hazy. This means that it is only in a preliminary stage and needs more thought. If it is not down on paper, it doesn't exist, at least from the point of view of the rest of the world.

If I still haven't convinced you that getting your idea in writing is critical, consider this: Writing down your idea at the very beginning of the grant-seeking process serves as a preliminary outline for your final proposal. Doing it now will make proposal writing that much easier later on.

Since the act of writing down your idea serves as an initial grant proposal, how careful should you be? There are two schools of thought on this issue: One recommends that you think of this writing as a generous act. Don't be too stingy in the first draft. Withhold suppression.

you can trim off the fat afterward. If you are too strict with yourself at this stage, you might edit out some element of potential importance.

The other point of view recommends that you avoid "first draftitis" or the habit of writing too loosely and carelessly thinking you can always fix it up later. This habit can become time-consuming, requiring too many subsequent revisions.

I personally suggest that you do it any way you can. If you have an idea, sit down right now and let the words flow from your pen. Although writing down ideas tends to limit them, it also helps to clarify them.

TESTING YOUR IDEA

Now that you have your idea in writing, the next step is to determine what to do with it. In the words of one inventor: "simply having an idea is a long, long way from doing something about it, and it's an even longer way from doing the *right* thing about it."[3(p.1)] In point of fact, most ideas work well on paper. How can you tell if your idea really will succeed if, indeed, you do receive a grant for it?

To help determine whether or not your idea is feasible, you might adopt some of the tests recommended for new inventions. Like an inventor, you can test your idea for desirability, value, and profitability (not necessarily in terms of dollars, but in terms of benefit to you, the funder, and/or society at large). Can it be made? How about a model, prototype, or real working example? What will it cost to produce? If at all possible, try out your idea on a limited scale on your own before you submit a grant proposal. This trial produces a "sales history" and nothing impresses a funder more than prior success. Positive feedback from members of the population who eventually will benefit from your idea will help beef up your "statement of need," an important component of your proposal to a funder.

REVISING YOUR IDEA

Once you begin to test your idea, you quite naturally will feel the need to revise it. The first tentative expression of your grant idea in writing may be a far cry from the finished concept you present to a funder in the form of a proposal. Don't be afraid of revision. Corrections should be viewed as opportunities for improvement, not as evidence of defects in you or your idea. The changes, additions, revisions, and refining your idea may undergo are many.

If revision is difficult for you, consider this list of "manipulative verbs" suggested by Koberg and Bagnall.[12] See if any of these verbs might be used to revise, expand upon, or otherwise improve your idea for a grant to make it a more attractive or more practical vehicle for funding.

Multiply	Distort	Fluff up	Extrude
Divide	Rotate	Bypass	Repel
Eliminate	Flatten	Add	Protect
Subdue	Squeeze	Subtract	Segregate
Invert	Complement	Lighten	Integrate
Separate	Submerge	Repeat	Symbolize
Transpose	Freeze	Thicken	Abstract
Unify	Soften	Stretch	Dissect

I recommend that you develop your idea horizontally, e.g., try different approaches, styles, and modes, as well as vertically, e.g., work out every detail each step along the way. Try to perform these two types of refining simultaneously since one stimulates the other. Concentrate on that one idea. Stay with it until you are satisfied. Add a little to it each day. You will rarely achieve success in one flash of inspiration. Above all, know when to quit. At some point, your idea will be fully developed and ready for the next step—selling it to a funder. Don't be afraid to follow your own instincts about when that will be.

Selling Your Idea

The first person you have to sell your idea to is yourself. Presumably, if you've gotten this far, you're pretty well sold on your idea already. If not—if you have doubts about the value, desirability, or practicality of your idea, you had better stop and reconsider whether or not you should seek a grant for it. If you are not sold on your own idea, you will not be able to sell it to others. If you are not purely and totally convinced that your idea is valuable and necessary and that you are the logical person to carry it out, you will never convince a funder to give you a grant. Remember, confidence is contagious; if you believe in yourself, others will believe in you.

The next group of people you will have to sell your idea to are those directly affected by your working on your project: family, friends, colleagues, co-workers, supervisors, and mentors. Without the support of these individuals who are important in your life, your work will seem twice as difficult. Sometimes, exposing your idea to those close to you is even more difficult than presenting it to funders, because rejection by those who know you may hurt more.

Consider the possibility of sharing your idea with family members, colleagues, co-workers, and supervisors for the purpose of premarket testing. Write down their comments, suggestions, and criticisms. Evaluate these recommendations as objectively as you can, keeping in mind the varying levels of expertise and familiarity with your subject matter, as well as the personal motivations of others for liking or disliking your idea. Next, revise your idea accordingly. If their comments are negative, remember that they are rejecting your idea, not you as a person. By the same token, don't allow yourself to be overly influenced by the opinions of outsiders. Remember, they might just be wrong. No one is as intimately familiar with your idea as you are.

Marketing is a process foreign to most grant seekers who are academics, students, artists, scientists, or inventors unschooled in the practices of buying and selling. As an incentive to this process, keep in mind that presenting your idea to a funder can be almost as creative a process as having the idea and working on it. Think of your idea as *attractive merchandise* and proceed accordingly. At first, the notion of viewing your idea as "merchandise" may seem vulgar. However, there is nothing dirty or unethical about packaging your idea in a way that will make it most attractive to others. Once you get the knack of presenting your idea this way, you will see that it is easy to market it without compromising yourself.

Protecting Your Idea

One final area of investigation is protecting your grant idea from others. There are two issues to consider regarding the protection of your idea—is it possible, and is it necessary?

In the words of Edward de Bono, "new ideas are everyone's business." An idea may originate with one individual. Yet the moment the idea holder seeks to implement his idea, it becomes no longer his personal property. In the same way, as soon as you, the grant seeker, turn your idea into a proposal, you can no longer expect to hold onto it securely. In a sense, the idea becomes the property of the funder, the population it will serve, and the public at large. You no longer own it.

If a funder doesn't misappropriate your idea, perhaps a colleague or even a friend or family member will. In fact, there are those, particularly in the field of invention, who recommend that you reveal your idea to no more persons than absolutely necessary. I disagree with this approach. The chances of having your idea for a grant stolen in most circumstances are small indeed—perhaps 1 in 1,000.[13] Even if you were to examine this one instance of theft, it would probably turn out that the idea holder discussed his idea widely with anyone who would listen but did nothing to implement it. Of course, if you wait long enough, someone else is bound to develop your idea.

If you discuss your idea selectively with those who have potentially useful know-how in your field and particularly in the area of funding for your field, there is a whole world of knowledge to be gained. Since you probably can't totally protect your idea anyway, it makes far more sense to share it judiciously with others who might be helpful to you in developing and implementing it.

On the other hand, if your idea for a grant involves an invention, product, or publication that is *legally* protectable (by patent, copyright, or other means), you should seek out legal or other professional counsel before discussing it with others.* Funders, like investors, are impressed

---

*If you're an artist, see *What Every Artist Should Know about Copyright* (New York: Volunteer Lawyers for the Arts, latest edition).

by your actions in this regard. I would like to suggest, however, that the very best protection you can provide for your idea is to *get on with it.*

IF DEVELOPING NEW IDEAS FOR GRANTS IS SO DIFFICULT, WHY BOTHER?

In the words of Rollo May, creative thinkers may be viewed as "frontier scouts who go ahead of the rest of us to explore the future."[14] The path to the generation and development of new ideas for grants is not an easy one. As already noted, it is fraught with difficulty, emotional conflict, and anxiety. The individual who has new ideas consequently feels himself to be simultaneously privileged and cursed. Yet despite the risks involved, the potential rewards in terms of self-development and benefit to one's fellow man make it all seem worthwhile.

POSTSCRIPT

I remember once hearing the following anecdote:

Henry Ford hired an efficiency expert to review the operations of his automobile plant. When the expert complained about one employee who sat all day with his feet up at an empty desk, Mr. Ford replied that several years earlier that employee had an idea that saved his company millions of dollars. Ford declared himself more than willing to pay the man's salary until the day he retired, just on the off chance that he might have another good idea.

NOTES

[1] Hooper, M. *Everyday inventions.* London: Angus & Robertson, 1972.
[2] Belcher, J. C., & Jacobsen, J. M. *A process for development of ideas.* Washington, D.C.: New Notes and Deadlines, 1976.
[3] Kracke, D., & Honkanen, R. *How to turn your idea into a million dollars.* New York: Doubleday, 1977.

[4]Groh, J. *The right to create.* Boston: Little, Brown, 1969.

[5]Adams, J. L. *Conceptual blockbusting.* San Francisco: Freeman, 1974.

[6]Kivenson, G. *The art and science of inventing.* New York: Van Nostrand Reinhold, 1977. P. 3.

[7]Heyn, E. V. *Fire of genius: Inventors of the past century.* New York: Doubleday, 1976. P. 158.

[8]de Bono, E. *Lateral thinking.* New York: Harper Colophon Books, 1970.

[9]de Bono, E. *New think.* New York: Basic Books, 1968.

[10]Winston, S. *Getting organized.* New York: Norton, 1978. P. 23.

[11]Greenhood, D. *The writer on his own.* Albuquerque: Univ. of New Mexico Press, 1971. P. 186.

[12]Koberg, D., & Bagnall, J. *The universal traveler,* Los Altos: William Kaufmann, Inc., 1981.

[13]McNair, E., & Schwenck, J. E. *How to become a successful inventor.* New York: Hastings House, 1973. P. 79.

[14]May, R. *The courage to create.* New York: Norton, 1975. P. 122.

" . . . a person's focus must be on his passion. He must be able to integrate who he is with what he is doing, see his project as a whole, and do his work systematically in order to legitimately expect the money to take on its secondary helping role."
—MICHAEL PHILLIPS
The Seven Laws of Money

Chapter 4 / GETTING INTO GEAR

ACHIEVING A STATE OF GRANT READINESS

In this chapter, I urge you to stop and reassess your desire for a grant as well as the grant idea itself before you proceed to seek outside funding. Many of the recommendations you will find here may seem like common sense. However, they are often overlooked in the rush to get to the proposal writing stage. In the words of Erich Fromm, "All our energy is spent for the purpose of getting what we want, and most people never question the premise of this activity: that they know their true wants."[1]

A prerequisite for getting into gear for grant seeking is achieving a state of readiness. Although the previous chapter was concerned with getting your idea into shape for funding, this chapter is concerned with getting *you* into proper shape, during what is commonly referred to as the preapplication phase of grant seeking.

As part of your effort to achieve a state of readiness, you should become aware of developing grant opportunities. Almost everything in your surroundings may be viewed as relating to your grant-seeking endeavor. Be alert to the mention of new grant programs. Attend meetings and conferences in your subject field; study professional journals; follow up on leads by writing for additional information; build up your own files.

It is vital for you to read materials relating both to your subject field and to the process of seeking grants. Visit special libraries and information centers that might have useful source books. Take courses, attend lectures, write to the experts. Contemplate your own prior experience and see how it might be relevant to your efforts to secure funding.

Aside from reading about your grant idea, it is important to talk about the idea as well as about your endeavor to seek a grant. As already noted, the chances of having your idea stolen are slim. The chances of its being enhanced by the opinions and suggestions of your colleagues and other experts are great. Reach out. Make contact with other individuals and organizations with similar interests. Share information and publicity along with praise and credit.

Also do not overlook the value of the "invisible network" as a source of much useful information and support as you approach a state of readiness. Communication by word of mouth spreads with extreme speed and spontaneity if the subject matter is close to the emotions of the people discussing it. Gossip at the local pub, faculty club, student union, or artists' hangout may be the source of information valuable to your grant-seeking efforts. Access to this invisible network helps you to become *au courant* in your field, a significant component of grant readiness.

There are several activities you can perform that will help build up a momentum:

- Keep a journal of your grant-seeking progress. Recording goofs as well as successes helps avoid repetition of mistakes and minimizes duplication of tasks you've already performed.
- Capitalize on local resources; look in your own back yard for help and support.
- Get on mailing lists of various funding and related institutions.

- Subscribe to relevant journals; build up your personal library and files.
- If possible, get to know some funding officials informally. Even if they can't fund you themselves, they can provide valuable insight into how grants or prizes are awarded.

YOUR OWN ATTITUDE IS ESSENTIAL

Getting ready to seek a grant is tantamount to developing the proper attitude, since, in the end, it will be your own personal attitude that will help determine whether or not your idea is funded. Grants decision makers like ideas that currently are in vogue and appear to be cost-effective. However, when funders support ideas, they are really backing the individual who will carry them out. What better way to judge an individual's merit than by his own view of himself? Your self-image or personal attitude as it will be portrayed in written and telephone communications, applications, proposals, and interviews, therefore, will be an important factor in the final decision whether or not to give you a grant.

Keep in mind that you are embarking on a professional relationship with the funder, not a supplicant-to-donor one. At the same time, do not be intimidated by funding institutions. Such organizations are operated by ordinary mortals like you. They probably have more money at their disposal, but you may have more in the way of talent and ideas.

To help develop a winning attitude as a prerequisite for grant readiness, begin with the positive. Sit down and make a list of everything you've got going for you at the outset of your grant-seeking endeavor. The positive attitudes that underly your proposal and that you will display in person, if you are lucky enough to secure an interview, will soon become apparent to the astute funding executive and may allay many of his fears about awarding you a grant.

Don't be afraid of appearing too aggressive. Grant seeking always involves asking someone for something. That someone may be more or less willing to give you what you want. However, first you must be willing to ask for it. Take the initiative. Don't sit back and wait to be chosen. Remember: "People who cut in are usually pretty good dancers."[2(p.42)]

Above all, don't give up before you ask. In the words of one foundation officer, "He who hesitates watches someone else get the grant he might have gotten."[3]

To nurture a positive attitude and to boost your own morale, peruse announcements of grants that have recently been awarded. Talk with successful grant recipients. Not surprisingly, the mere knowledge that someone else got a grant will spur you on to work even harder.

PRESCRIPTION FOR A POSITIVE ATTITUDE

In preparation for seeking a grant, there are five rules you can adopt to enable you to have a more winning attitude:

1. Be a friend to yourself; act in your own best interests. Reward yourself. Compliment yourself on a job well done. Blow your own horn.
2. Associate with people who are confident and those who already have achieved some modicum of success.
3. Keep busy. Tackle the things you most dislike. First confront the task you dread most of all. Such confrontation defuses self-doubt.
4. Fight performance anxiety (which can be viewed either as fear of failure or of success). To help yourself combat anxiety, exaggerate in your mind the situation you most fear until it seems humorous or absurd. For example, if sitting down to write a proposal frightens you, imagine yourself chained to the desk. If the prospect of an interview with a funding executive intimidates you, picture the two of you trapped together for hours in a stalled elevator. If receiving a letter of rejection seems scary, imagine the mailman arriving in executioner's garb with a sword in his hand.
5. Finally, set out to do what you really want to do. You are better off trying to please yourself because you can't please everyone anyhow, no matter how hard you try.[4(p.63)]

PROCRASTINATION

As a footnote to this discussion of grant readiness and attitude, I cannot fail to mention procrastination, since this is a subject that plagues every grant seeker, be he novice or accomplished fund raiser. Procrastination has been defined as "the habit of putting off until tomorrow what is uncomfortable to do today," and we all suffer from its ill effects. Procrastination in grant seeking thwarts both present and future success since the inevitable result is that opportunities for seeking funding are irretrievably lost.

In his book, *Do It Now,* Dr. William J. Knaus[4] recommends some antiprocrastination techniques or "time saving strategies" that you can employ to get your work under way:

The "five minute plan." Set a finite period of time on a steady basis to work on your grant seeking. Write fixed appointments with yourself in your calendar and be sure to keep them.

The "bits and pieces approach." Tackle one small aspect related to seeking a grant every day. In this way, the work seems less overwhelming.

Concentrate on one priority at a time. Avoid the feeling of being swamped, which comes from believing that all tasks are equally important.

To the strategies outlined by Knaus I would add:

Allow enough time for seeking grants. Avoid overloading yourself by underestimating how long a particular task will take. Permitting yourself sufficient lead time is particularly important for the grant seeker, because seeking funding is a time-consuming process. If you desperately need money next week or even next month, you probably should try to find some way other than a grant to support your idea, at least for the time being. The typical grant-seeking endeavor may take at least three months or up to nine months of hard work. It may be almost a year from the time you apply before you actually are notified by a funder whether or not you'll receive a grant.

To protect yourself against the tendency to procrastinate, take note of the successful grant seeker's motto: Plan ahead and observe deadlines.

If after adopting these techniques you still experience an overwhelming resistance in yourself to the process of seeking a grant, face it head on. Ask yourself—What does this resistance mean? Should you abandon the project entirely or just for now? If you do decide to drop it, really do so. Don't let guilt clutter your mind. If you find you can't just let your grant idea go, see if you can hire someone else to do the distasteful parts for you or delegate them out to others in exchange for services. At the same time, be sure to promise yourself small rewards for segments completed.[5]

THE IMPORTANCE OF PLANNING

Like attitude, planning is essential to a state of readiness as a precursor to successful grant seeking. In order for planning to be effective, it must be systematic. Create a grant-seeking timetable for yourself and adhere to it. On a single sheet of paper, write down projected start-up and completion dates for each component of your grant-seeking endeavor, keeping in mind that some may overlap. Remember to be generous in the allotment of time for yourself to accomplish each stage.

Ask yourself, are there legal steps you should take prior to applying for a grant? Are permissions or releases required? Do any official agencies have to be formally notified? Is the approval of your immediate supervisor necessary? All of these questions fit under the rubric of advance planning. Evidence of sound planning is one measure applied by grants decision makers who will evaluate your proposal.

As part of your systematic planning, think of all the possible obstacles to the success of your grant search, such as nonavailability of critical data or the reluctance of crucial people to cooperate. Are there likely to be strong objections to your grant idea? Who will raise them? How will you counter them? Is there a risk of partial or complete failure if, indeed, you do receive a grant?

The next step is to try to figure out what changes might come about if your grant idea is funded. Determine to the best of your ability what external events, beyond your control, may have an impact on your idea and its development. Be flexible and prepared to modify your plans if resistance is met or unforeseen developments occur. If problems arise, be ready to try for at least a partial victory.

Financial planning is also an important part of grant readiness. How much to ask for represents a thorny question for the novice grant seeker. To help justify the amount you will eventually request, from this moment on you should keep track of *all* expenses, no matter how small, related to the development of your grant idea. Such documentation will come in handy later when you construct an actual budget for your grant idea as part of your proposal. (See Chapter 7.) "Putting a price tag on your lifestyle"[2(p.109)] is a difficult but necessary element of getting ready to seek a grant.

SPECIFIC RESOURCES FOR THE GRANT SEEKER

As part of your advance planning to get ready to seek a grant you may wish to make contact with the following individuals or groups:

• Look for successful grant recipients—leaders of organizations and individuals. Use current grants lists, press, and other announcements to identify them. Write or call and ask how they did it. You may be surprised how generous they will be with their time. Speak with colleagues and other grant seekers. Although they may know no more than you, such discussion helps stimulate the creative juices.

• Solicit advice from experts or those renowned in your field. Talk to them about the state of funding in your area. Ask their opinion of your chances for funding. See if they can put you in touch with other experts.

• Make a list of whom will be most affected by your idea if it is funded. Share the idea with some of these individuals to get their reaction.

• Contact your Congressman or other local public official. See what this official can do for you. Ask what useful information his staff has at hand. Your Congressman or Senator may have a grants person assigned to his office (or some staff member with similar responsibility). Duties might include writing letters of endorsement, securing and providing application forms, helping to prepare proposals, and making contact with appropriate government agencies. This staff member may be a fount of knowledge and a highly useful contact for the individual seeking a grant.

• If you are affiliated with a sponsoring institution, get in touch with its office of development, if it has one. This office often prepares proposals and budgets and provides backup services. Their staff may even help identify potential funders. Clearing your grant idea with the development office, if it exists, and with the leadership of your sponsoring organization, if there's no development office, are indispensable preliminary steps to grant readiness.

• Educate yourself. There are independent groups, such as The Grantsmanship Center in Los Angeles, that offer free or low-cost workshops, consultations, training, proposal evaluation, and management assistance to individuals or to leaders of nonprofit organizations seeking grants. Various professional associations hold conferences, and some universities have courses related to grant seeking. If you have a sponsor or employer, you may be able to get this organization to pay for you to attend such a training session or course. Before enrolling in a workshop or course, ask for names of grant seekers who have already taken the class to see how they rate it

• Secure free advice or paid assistance when you need it. Professionals exist, including fund raisers, lawyers, bookkeepers, accountants, and public relations people, who can help you if necessary.

• Touch base with the leaders and staff of local community groups. This type of contact is good public relations and may help you gain important leads.

• If you're an academic, members of the faculty in other universities as well as your own are an excellent source of grapevine information.

• If you're an artist, each state has its own arts council or commission, which not only sponsors grants, fellowships, or short-term residencies but often helps individuals apply for outside funds.

• If you want to study or travel abroad, contact the cultural attaché of the country in question, usually at its embassy in Washington, D.C.

I am sure, if you think about it, you can come up with your own additions to this list of resources to help you achieve grant readiness. The important thing to realize is that such resources exist, enabling you to reach out and ask for help, thereby creating your own advocate network.

GETTING YOUR OWN HOUSE IN ORDER

One final element in getting ready to seek a grant, just as important as adopting the proper attitude, establishing a systematic plan for action, and making contact with appropriate resource people and organizations, is the creation of a supportive environment for the work on your grant-seeking endeavor. Whether you work in a business office, library, or at home, having sufficient backup records, handy files, and consistent book-keeping procedures will greatly facilitate your effort to seek a grant and should save you substantial work time. If you can figure out in advance what documents, reports, reference tools, and so on you will most often refer to, and find a way to make these readily available, you will also cut down on your work time. The following sections provide brief suggestions on references, endorsements, resumes, filing systems, and a basic personal reference library for the grant seeker.

REFERENCES

Just as there is no one without connections, none of us is truly "unknown." You may not be famous in your field. You may be just starting out, or your project may involve completely untried ideas. Nevertheless, somewhere there is an employer, colleague, or mentor who himself has a good reputation and will provide you with a positive reference.

Important suggestions to keep in mind about references are that you (1) select people who know you well, (2) line them up *in advance* of seeking a grant, and (3) warn them that they may be called upon to provide a reference for you. It is better to choose people who really know you and can be counted on to give you praise rather than well-known luminaries who barely recognize your name, although you were once their student. Also, make sure that those whose names you give as references know about your *present* work and can comment upon it in an intelligent way.

Your references may be no more than a list of several names attached to your proposal. They are important, nevertheless. If your idea for a grant receives serious consideration by a funder, it is a safe bet that funding executives or their staffs will contact those whose names you've given as references. This is one aspect of the grants decision-making process over which you have a modicum of control.

ENDORSEMENTS

Endorsements are expressions of support and approval from famous, influential, or authoritative individuals or groups. They may be merely touched upon in your proposal or mentioned in your cover letter (see Chapter 7). Securing endorsements, however, is an important part of getting into gear for grant seeking, because endorsements help you build community support and acceptance for your grant idea.

References are used to enhance your credibility as an applicant. Endorsements relate more to the benefits of and need for your idea. Unlike references, endorsements need not come from those who know you or your work well. It is advisable to find "name" organizations or institutions to endorse your ideas, whenever possible. The more prestigious and influential the endorser, the better.

As with references, be sure to have endorsements lined up in advance of applying for a grant. Common sense rules of courtesy apply to endorsements. Ask for them early since, especially in the case of nonprofit organizations, board approval may be required. Request that the endorsement be given in writing, on official stationery if possible, because word of mouth usually is not enough.

Always send a thank-you note to the endorser.

You must *ask* for endorsements, since they seldom are given gratuitously. However, be sure when you do ask to ascertain if some action on your part is expected in return. Above all, be certain of the endorser's commitment. Lukewarm endorsements may do more harm than good and are to be avoided.

Whom should you ask to endorse you? Local community agencies, groups who might directly benefit from your project, state or local officials, heads of prestigious nonprofit, educational, or research bodies, colleagues, even members of the competition all might provide endorsements if they like your idea. If you have a sponsor or employer, the leaders of that organization should give your grant idea their blessings before you proceed.

RESUMES

There have been reams of material published about the resume, some of which I'm sure you've already read. Consequently, I will simply touch upon the resume here, primarily as it relates to the individual grant seeker.

Writing or rewriting your resume is an unpleasant but mandatory task to be undertaken prior to seeking a grant or award as well as prior to seeking employment. It is unlikely, however, that the same exact resume you used to seek a job will be appropriate for obtaining a grant. Although there are some similarities, the two processes are actually quite different.

The entire thrust of your grant-seeking resume should prove that (1) you are the most logical person to carry out your grant idea; (2) you are mature, responsible, and capable of success; and (3) you have the proper educational background, training, and professional connections to carry out your idea successfully. Bearing these three axioms in mind, reexamine your present resume. If it does not portray you in the most favorable light as a potential grant recipient, then draw up a new one.

The resume has been dubbed "that most formidable of personal documents."[2(p.157)] The process of creating a new one is indeed intimidating. You must, nevertheless, overcome your aversion and begin. Here are several rules of resume writing for grant seekers of which you should be aware:

1. Never say or imply anything negative about yourself in writing.
2. Never lie or exaggerate—although there's nothing wrong with presenting your qualifications in the best possible light.
3. Be selective about the things you put down. For example, if you are applying for a grant to hold a photographic exhibit, potential funders do not need to know where you went to elementary school.

4. If you've ever received a prior grant or award, be sure to note it.
5. Be certain that each item on your resume is up-to-date, right up to the present moment, before you apply for a grant.

FILING SYSTEMS

Getting your house in order—setting up an accessible area for your files and records—facilitates grant readiness. Neatness, the ability to organize, and coherence of thought are all traits to be developed by the grant seeker, since ultimately they will contribute to your getting a grant.

Keeping good records is sometimes tedious and may seem to prolong your work time. It is, however, a vital operation in the successful grant effort. Hence, you should try to develop a record-keeping system that will lighten the burden of seeking a grant.

Filing systems are difficult to write about because they are idiosyncratic. Straight alphabetical arrangements work better, in general, than subject or other classification systems, because they are more straightforward and flexible. Separate files should be made up for each potential funder. You should also have a file on "new ideas for future grant projects." In addition, it is handy to have a "credibility" or "kudos" file, including news clippings, promotional literature, samples of your work, letters of endorsement, and awards received. To ensure the efficient operation of your grant-seeking endeavor, it is also advantageous to have "things to do," "to file," and "pending" folders, as well as a separate address book or Rolodex for useful personal contacts.

Good filing systems require steady maintenance in order to remain up-to-date. Each time you pick up a folder, rifle through it quickly to pick out and discard the dead wood. The decisions as to what to keep and what to discard almost always present problems for the novice. In general, you should retain only those materials that appear to have specific value and contribute, enhance, or support your idea for a grant. If a piece of paper does not meet these criteria, throw it out!

For those items you decide to keep, what headings should you use? Pick out the most obvious key word that seems logical and memorable to you. Then file alphabetically, using the key words you've selected as headings.

BASIC PERSONAL REFERENCE LIBRARY

To make the work on your grant idea easier on yourself, store all pertinent publications and materials together in one location near the desk or table where you work. The kinds of publications you might want in this vicinity are proposal writing guides, fund raising guides, directories of government, corporate and foundation funders, indexes of grants awarded by those funders in your subject field, directories of funders in your locale, periodicals relating to grants and to your subject field, newsletters, announcements, and annual reports and other informational leaflets issued by funders. Publications falling into these categories and others will be discussed in detail in Chapter 6 and evaluated as to their usefulness for individual grant seekers.

The question of whether you should buy any of these materials or simply plan to use them at a local library depends upon your budget. Many of the standard reference books are quite costly, but some newsletters, periodicals, and annual reports are free or relatively inexpensive. If you refer to a publication daily or even weekly, *and* you (or your sponsor) can afford it, you should probably own it. If not, seek out public, academic, or special libraries in your locale that have the materials you need. Plan to spend some time in the library since most of these kinds of books are considered reference works and, therefore, will not be available for circulation.

No matter how small your budget, I recommend that you have at least a good pocket dictionary, style guide, thesaurus, and up-to-date local telephone directory ready at your desk as you begin your grant-seeking endeavor.

To sum up this discussion of getting into gear, advance planning is the sum and substance of getting ready for successful grant seeking. Adopting the proper attitude is also essential since it has a significant impact on the decisions funders will make concerning your proposal. Getting ready to seek a grant requires you to identify your strongest skills and resources, keep your options open and avoid restrictive thinking, and construct realistic timetables for yourself. All of these processes are time-consuming but necessary, and ultimately very rewarding, since seeking a grant obliges you to get your life in order.

NOTES

[1] Fromm, E. *Escape from freedom.* New York: Holt, Rinehart & Winston, 1976. P. 252.

[2] Bolles, R. N. *What color is your parachute?* (Rev. ed.) Berkeley, Calif.: Ten Speed Press, 1978.

[3] Hartwig, J. C. Betting on foundations. In *Foundation grants to individuals* New York: The Foundation Center, 1977.

[4] Knaus, W. J. *Do it now!* Englewood Cliffs, N.J.: Prentice–Hall, 1979.

[5] Winston, S. *Getting organized.* New York: Norton, 1978. P. 48.

"People give to people."

—Lawrence Lee
The Grants Game

When the Director of the Aeolians, a chamber music group which receives government and foundation grants, was asked by a New York Times *reporter whether or not the group had been affected by federal budget cuts, he quipped: "Only financially."*

—*New York Times*
February 7, 1982

Chapter 5 / FACTS ABOUT FUNDERS

WHY KNOW YOUR FUNDERS?

One of the basic tenets of successful grantsmanship is being well informed. To the individual grant seeker this means, above all else, knowing your funder. For you cannot have too much information about a particular funder. The more you know, the greater the likelihood that you will be able to write a tailor-made proposal for each specific source of funding. This means having hard, cold facts at your disposal, not half-baked information, rumors, or hearsay.

Yet in my own contact with grant seekers, it has become apparent that many have little or no idea about the standard operating procedures followed by those who give money away. The typical grant seeker often doesn't know what funders really do, how they operate, or what their priorities are. Worse yet, many grant seekers haven't even considered these questions. All too often, they are so absorbed with their own need for money that they haven't had time to realize that funders too have goals their managers strive to accomplish. This lack of consideration is a sadly inappropriate and ill-prepared way to begin seeking funds.

In your own search for a grant, you can help remedy the situation I've just described by banishing all stereotypes relating to funders from your mind. Rather, you should find out the facts. Next, stop thinking of funding agencies as institutions. They are composed of people, human beings who have elected to do this type of work, individuals with needs and desires similar to your own. The more you know about them, the more similarities you can draw between their situations and your own. Hence, the better your chances of getting a grant.

Keep in mind that getting funded is merely a vehicle to enable you to accomplish your real goal—to get on with the work on your grant idea. The strongest argument I can make for knowing your funder well is that it enables you to start out with realistic expectations. These in turn help you save significant amounts of time—a boon to you, since you want the actual processes of finding and securing a grant to take as little time as possible away from your precious work.

Beginning with realistic expectations helps you in three important ways:

1. It eliminates false leads to inappropriate funders.
2. It cuts down on grandiose presumptions about the amount of money you can expect to receive. (*No one* gets rich on a grant. Such funding usually covers just the bare bones of facilitating your idea plus direct work-related costs.)
3. It enables you in advance to project a fairly accurate estimate of the amount of time the entire grant seeking process will take.

To provide yourself with realistic expectations, you should start out grant seeking with as much raw data as possible. Specifically, for each funder you intend to approach, you will need answers to the following questions: What current trends exist, resulting in limitations, restrictions, and/or potential future opportunities for individual applicants? Although no one can provide you with guarantees, how good are your chances? Try to find out about the competition: Who was awarded funding in the latest year of record? Might they be reapplying this year? What percentage of applicants do these grant recipients represent? Who are the important decision makers employed by the funder? What review process do they follow? How long will it take? When will you be notified about their decision regarding your proposal? What style of approach should you adopt? What politics and/or proper etiquette are involved?

The resources and methodology for finding out the precise answers to many of these questions as they relate to specific potential funders are discussed in detail in Chapter 6. These questions are also dealt with in a more general sense, in terms of trends in grant making as they relate to various categories of funders, at the start of each section in this chapter.

Why Facts and Figures?

The facts and figures presented in this chapter are not just window dressing. The serious grant seeker should become familiar with the basic factors affecting grant making, including both limitations and possibilities set forth by each category of funder. Becoming thoroughly familiar with funders may enable you to create your own grant opportunities even where presently none seem to exist. Finding out about current restrictions and exclusions helps the individual applicant to discover ways to work around them.

Of course, no one expects you to memorize all the facts provided in this chapter. Rather, they are included here to help you become aware that, in effect, by considering your funder as well as yourself, you *are* considering yourself, for no funder exists in a vacuum. There are key events in the history of particular funders as well as present pressures upon them that either directly or indirectly affect your chances. Understanding and working with these elements helps you to get a grant. Being considerate of funding decision makers and knowing what to expect (e.g., avoiding surprises) are indispensable guidelines for successful grant seeking.

INTRODUCTION TO FUNDERS

In this chapter we will look at five categories of funders:

Foundations. Nonprofit, privately funded organizations that make grants for charitable and other purposes (a more precise definition is provided at the beginning of the foundations section).

Government. Federal, state, and local agencies which make grants or other monies available to U.S. citizens.

Corporations. Contributions programs of the American business community.

Individuals. Donors who make gifts to various recipients during their own lifetimes and/or by bequest.

The miscellaneous category of funders. Those hard to define associations, clubs, local groups, and agencies that often make small cash gifts, prizes, or awards to individuals.

THE ABSENCE OF HARD DATA

Each year the American Association of Fund Raising Counsel* compiles detailed statistics on charitable donations to nonprofit organizations. By contrast, no precise, comprehensive statistics have ever been compiled for private or governmental support of individuals. Historically, this subject has been largely ignored by statisticians, researchers, and therorists active in the field of philanthropy. No one really knows, therefore, precisely how many grants or how much money is awarded either directly to individuals or via sponsoring arrangements by the various categories of funders. At the same time, we do know that grants to individuals *are* awarded each year by certain funders on a more or less regular basis. An educated guess would be that they comprise a tiny percentage (perhaps less than 5%) of the total philanthropic picture, but still a significant amount in terms of real dollars to the individuals who actually receive the grants.

---

*The AAFRC is a membership organization of fund-raising counseling firms which studies and identifies economic and social trends in American philanthropy.

The grant seeker should not be discouraged by the absence of hard data on grants to individuals. As noted elsewhere in this book, many grants awarded to organizations are actually earmarked for individual recipients who have fairly loose ties to their "sponsoring" organizations. There are, furthermore, ways to get around many of the limitations and restrictions imposed by funders on their grants to individuals. The opportunities exist. It is up to you, the grant seeker, to learn how to best approach funders in order to make it easy for them to give you a grant. As we shall see, grants to individuals generally are awarded under a completely different set of rules than those governing grants to organizations.

GENERALIZATIONS, MYTHS, AND STEREOTYPES

During the course of this chapter some generalizations about the various categories of funders are unavoidable, if we are to examine their similarities and differences. Stereotypical thinking, nonetheless, is discouraged, and I strongly advise that you approach each funder on as individualized and personal a basis as possible. (Treat each one as if it were an exception to the rule.)

Myths about funders are well known to grant seekers. Some of the following widely held beliefs no doubt already sound familiar to you:

- Foundation administrators are secretive to the point of paranoia. They don't give to individuals. Most foundations are just tax dodges anyhow.
- Corporate givers solely are concerned with profits. Corporate grants decision makers support only their own employees and relatives of owners or managers. Most corporate philanthropy is closely tied to public relations activities and advertising.
- Federal government funding agencies are essentially bureaucratic. There are so many forms to fill out and so much red tape that the individual with the most tolerance for frustration and delay gets the grant.
- Political pressure prevails when it comes to state and local government funders. With increased competition for fewer government funds the individual applicant is bound to lose out.

- Wealthy individual donors exist who could help you if only you knew how to reach them. There are eccentric billionaries to whom a substantial gift to you would represent merely a drop in the bucket.
- There are associations, societies, clubs, research institutes, and so on, to benefit every imaginable subject field and category of recipient. These groups often have substantial sums of money at their disposal and operate at the whim of their chief executives. There are virtually no rules governing the awards, honors, and prizes these "miscellaneous category" funders bestow.

Are these all too familiar statements true or not? The following pages will tell.

GOVERNMENT FUNDERS

In a time of huge federal budget cutbacks and wildly fluctuating circumstances relating to government grant expenditures, it is difficult to discuss the subject of government grants to individuals except in a very general way. By the time this book goes to press, funds may be more severely cut back, presently existing programs may be obsolete with new ones formed, and the entire political climate probably will be changed to some degree. It is possible, however, to make some generalizations that pertain to most government granting agencies, be they federal, state, or local, *at the present time* and to remind you that with government funders, perhaps more than others, it is essential to check and double check to make sure your information is up-to-date. Pick up the telephone, contact the appropriate office, and ask for the most current guidelines and for the next date of application. Unless you constantly take steps to protect yourself, you run the greatest risk of wasting time when applying to government funders.

FEDERAL FUNDING AGENCIES

No one really knows precisely how much the federal government gives away each year (since there are so many different ways of tabulating these figures). Yet there is general agreement that the total figure in recent years has been at least $100 billion for some 1,400 grant and contract programs, $10 to $30 billion of which went to individual and independent nonprofit organizations. All of these expenditures make the U.S. Government the biggest single source of financial aid today, and competition for these programs is steadily increasing.

During 1981, appropriations for the National Endowment for the Arts (NEA) were approximately $159 million; appropriations for the National Endowment for the Humanities were approximately $151 million. Even though they represent a miniscule amount of money in relation to the total federal budget, these appropriations were substantially cut by the end of 1982, but not by the 50% figure initially proposed. The grants to individuals programs of the National Endowments and the National Science Foundation will be discussed in detail later on in this section.

In addition to the "government foundations"* many other federal government agencies operate programs of possible benefit to individual grant seekers. To help identify these programs see the current *Catalog of Federal Domestic Assistance* (discussed in Chapter 6). Here are some examples of the wide variety of federal programs providing opportunities for individuals.

- The Small Business Administration (jointly with NEA) finances a national program of seminars to teach business and marketing skills to artists.
- The Appropriate Energy Technology Project offers two-year grants of up to $50,000 for new ideas on energy conservation techniques. Approximately one-half of these grants go directly to individual inventors.
- The National Institutes of Health have a program of research service awards for scientists that provides funding for full-time research training of individuals. Somewhat like loans, these

---

*The National Science Foundation and the National Endowments are called *government foundations* because, although officially part of the federal government (they were set up by Congress and receive their annual appropriations via this body), their grant-making operations strongly resemble those of the large, national private foundations.

awards include specific service time or payback provisions on the parts of the recipients.

- The Office of Indian Education operates a fellowship program for American Indian students in business administration, engineering, natural resources, education, law, and medicine, whereby full tuition, stipends, and other expenses are provided for up to four years.
- The Smithsonian Institution, the National Gallery, and the National Trust for Historic Preservation all have funding programs for research, training, and/or professional development of individual specialists and scholars in the social sciences and several other disciplines.

The above examples of federal government grant programs are listed here for the diversity they represent rather than for any consistent pattern of federal funding one can draw from them. Like the Supplemental Opportunity Grants, college work/study, and student loan programs, they may be terminated or severely cut back, or their future may be imperiled under federal budget cutbacks instituted by President Reagan. However, if the past record is any predictor, other funding programs will crop up to take their places, perhaps via state and local government agencies, acting either as intermediaries or as the actual disbursers of grant dollars.

STATE AND LOCAL FUNDERS

Although the government grant picture has been dominated up until now by the federal government, it is a safe bet that over the next ten years state and local funders will gain dramatically in importance. All of the states presently receive money from the federal government that they in turn dispense in the form of social and grant programs. Many have additional appropriations from their own legislatures. Some states even operate their own programs of direct assistance to individual grant seekers. The New York State Creative Artists Public Service Program is an example of one of these.

As a grant seeker, one rule of thumb to keep in mind about state and local funders is that if you've found a potentially good grant source, change is probably in the offing. As soon as you get used to dealing with one agency, following one style of approach, or filling out one particular application form, another will appear to take its place. Finding out about state and local funders and keeping on top of the many changes going on is discussed in the next chapter. When applying to local funders, political contacts you may have never hurt. In fact, the more politically connected people you know, the better. Also, expect many strings attached to state and local level government grants. For example, you may be expected to give lectures, conduct readings, work on community projects, provide publicity, and so on, in exchange for the funding.

Some generalizations may be made about government funding agencies, be they federal, state, or local, which tend to hold true no matter what the political climate is or the amount of grant dollars involved. Most government grant programs are mission rather than research oriented; hence, their results should be directly measurable. Government grant decision makers tend to prefer efforts that will produce definitive models that others may copy or adapt. Also, these funding executives generally respond quickly and negatively to controversy. This is especially true of bureaucrats at local agencies. No government funder will commit to indefinite periods of support.

For individual grant seekers, applying to government funders, at whatever level, presents a true challenge, since there is intense competition for a shrinking number of grants. Because government grant and contract seeking is a full-time, continuous process, requiring frequent shifts in tactics, the individual with limited time, few specialized resources, and little or no backup services is at a distinct disadvantage when compared with the nonprofit organization or contract-seeking corporation. Many government grant programs carry stiff and rather burdensome management and reporting requirements; all are subject to audits, resulting in much red tape for the grant recipient. The unaffiliated grant seeker, furthermore, may encounter some difficulties when applying to local funders since most require "eligible host organizations." Although these facts represent the discouraging news about government grants, the grant seeker should be encouraged to hear that at least one expert on grant giving (Waldemar A. Nielsen in a *New York Times* article) expresses the belief that the federal government has its most significant role yet to play in the funding of individuals.[1]

In summary, I would offer the government grant seeker the following words of advice: Do not apply for government grants unless you are willing to play the game by the funders' rules. You must have a certain tolerance for frustration and a willingness to confront red tape. Patience is an absolute necessity to the government grant seeker. You should be prepared to keep careful records, submit reports promptly, and expose your project to the scrutiny of outsiders. Although it is not necessary to sell your soul in order to receive a government grant, you must realize that in some ways your ideas will become no longer your own personal property. As a potential recipient of public funds, you will be held accountable to the public for your actions. (More on the responsibilities of the recipient of government funds will be found in Chapter 8.)

CONTROVERSY OVER PUBLIC SUPPORT OF INDIVIDUALS

Waldemar Nielsen's assertion that the federal government has its most significant role yet to play in the funding of individuals perhaps seems overly optimistic in light of recent developments regarding student loans, food stamps, the school lunch program, and others. This brings us to the examination of one issue of potentially significant impact on the future of government funding: There is, and has been for some time, strong feeling in certain sectors against *any* direct government support of individuals. This is no cause for alarm among grant seekers. In the worst case, you may be required to go through some intermediary agency (state, local, or independent nonprofit) to qualify for funding. However, becoming aware of the controversy surrounding government grants to individuals and the arguments on both sides of the issue helps you to better understand the reasons government funders choose to give or not to give and provides you with ammunition to aid in the presentation of the case for your own grant idea to a potential funder.

The arguments against government grants to individuals go something like this: Public support of individuals promotes self-indulgence and waste. It encourages recipients to be irresponsible. One British writer who favors withdrawing all public money from artists speaks for many critics of grants to individuals when he points out that since such grants are likely to be pretty small, they are not carefully scrutinized, and the recipient tends to spend the grant money too quickly and in too cavalier a manner.[2]

Some believe that grants to individuals tend to be "make work"-type awards for which the recipients, a number of whom are members of minority groups, are less than qualified.[3(p.59)] Michael Straight, former deputy chariman of the NEA, is himself opposed to federal government grants to individuals. He sees them as handouts and feels that the federal government lacks the wisdom and capacity to make good individual awards.

Straight as well as others with similar views favor the decentralization of the patronage power currently held by the government foundations and a select group of federal granting agencies. Joseph Papp, head of the New York Shakespeare Festival, went so far as to recommend that the Endowments be abolished, with federal funds rechanneled through state arts councils. The philosophy behind this recommendation is that important decisions about grants that will have substantial local impact should be made closer to home. There was even a bill proposed by the Senate Appropriations Subcommittee on the Interior (and eventually abandoned) that dictated that all NEA fellowships to individuals be eliminated in favor of block grants to state agencies and matching grant programs.

Much of this criticism of the federal government's handling of individual grant programs may be viewed as part of a backlash effect. Senator William Proxmire with his "Golden Fleece Awards" may have frightened some government funding executives to such an extent that henceforth they will fund only the most pedestrian of projects. Straight asserts that his recommendations about phasing out fellowship programs are made in order to protect NEA from further criticism. In his words, "the $5,000 awarded to Erica Jong (to write *Fear of Flying*) might well have imperiled the millions of dollars each year to museums, symphony orchestras, and opera, dance, and theatre companies."[4] Like Straight, other proponents of continuing federal support of the arts, scientific research, and social programs fear that a few grants to controversial individuals could attract enough attention to put the entire future of federal funding in doubt.

Yet there are many arguments to be made in favor of direct federal government grants to individuals. The most obvious of these is that the entire community benefits from the work of individual grant recipients, few of whom earn so much as a living wage from federal dollars. Since federal funds tend to incite private funding, cutbacks could well bring about concomitant private cutbacks, ultimately resulting in fewer funders making grant awards at all. To quote Harold Schonberg in an article about New York state's share of arts and humanities grants: "If nothing else, government support is a kind of seal of approval that generates a great deal of private money. . . ."[3(p.58)]

When the endowments and other federal granting agencies were originally formed, direct funding of individuals was viewed as one of their legitimate functions. A *New York Times* editorial expressed the fear that the aims and standards outlined in the original legislation that set up the Endowments have become steadily compromised and that Endowment and other federal grant funds increasingly have been used to serve social and political policy.[5] The *Times* editors urge the return to more grants to artists, writers, composers, and scholars in order to refocus the Endowments on the pursuit of excellence. In their view, grants made directly to individuals tend to be far less politically motivated than those to organizations or groups, hence, fairer and more impartially awarded to better qualified recipients.

Finally, one cannot help but ask the question: If the federal government stops making grants to individuals, who will pick up the slack? Who will take the risk on an unknown invention, untried vaccine, or new work of art? If such funding were to cease, society would be the great loser.

GOVERNMENT FUNDING AS PART OF THE TOTAL GRANTS PICTURE

The future of direct federal funding is in doubt. Certain select demonstration grants continue, but ongoing federally funded projects are fast terminating, and operational grants are on the decline. Coming to a close is the heyday of the categorical grant or program with federal dollars set aside for broad fields of endeavor like those cited earlier. Instead, block grants have come into favor. Block grants refer to money from the federal government that is passed down to state or local units according to rather complicated formulas based on such factors as per capita income, population, and taxation. Then the state or local agencies administer the funds.

This process may be good news to you, the grant seeker, because it provides more opportunity for direct contact with funders at the local level. In theory, at least, the grant ideas of individual applicants need have only a projected local impact to qualify for funding, as opposed to the national or regional impact needed to qualify for funding under categorical grants.

However, along with decentralized control of federal funds comes greater susceptibility to political influence. Most individual grant applicants do not have the kind of political clout needed to compete. The combined effects of decentralization and federal budget cuts could be potentially devastating to many local nonprofit organizations and individual grant seekers, unless they band together and take political action to open up the funding process to the broadest possible base. At the same time, grant seekers must become familiar with behind-the-scenes procedures at the state and local funding level and learn to cultivate influential contacts.

The *new federalism* is a term used to categorize the decentralization process just described. Before we begin our discussion of the new federalism as it relates to grants to individuals, a capsulized version of the history of U.S. philanthropy might be instructive as background information.*

U.S. Philanthropy: A Capsulized Version

In colonial times and throughout the 18th century, direct person-to-person giving was deemed necessary and greatly encouraged by society

---

*For further details on the subject, see F. Emerson Andrews, *Philanthropy in the U.S.: History and Structure* (New York: The Foundation Center, 1978) and Edward C. Jenkins, *Philanthropy in America* (New York: Association Press, 1950).

at large, and the church, in particular. By the 19th century, secular agencies were formed and national organizations were established, with local chapters to meet the welfare and health needs of the poor. The new profession of social work developed and with it the birth of truly institutionalized philanthropy as we know it today.

During World War I, mass fund-raising campaigns, like the strikingly successful ones conducted by the American Red Cross, began. This period also marked the beginning of extensive corporate donations along with the first instances of joint solicitation of funds and federated giving. Characteristic of U.S. philanthropy is the fact that gifts most often pass from the donor through middlemen, composed of various voluntary agencies, before reaching the intended recipients.

During the depression years, there were many highly publicized instances of one individual giving to another. At this point in our history, private welfare agencies were found to be inadequate. In 1935, the Social Security Act was passed, for the first time bringing under the province of government many services formerly charged to the private sector. Foundations were being formed and proliferating in the thirties. Also in 1935, the Revenue Act was passed, which authorized corporations to deduct (with a federal stamp of approval) up to 5% of their net income for charitable contributions.

In the years after World War II, 90% of the U.S. foundations were formed. This growth led to congressional scrutiny in the fifties and sixties and eventually to the Tax Reform Act of 1969, which, among other things, prohibits certain activities and imposes an excise tax on foundations, as well as strict rules regarding the awarding of grants to individuals.

During the Kennedy and Johnson years, it was assumed that government services could provide for society's elemental needs and private funders could thus turn their attention to more creative, e.g., esoteric, functions. However, today our confidence in government's ability to satisfy basic human needs is indeed shaken. Our economy is in a dangerous inflationary state, our stock market is unstable, and talk of "recession" and even "deflation" is commonplace. All of this economic insecurity has a grave impact on grant making, both public and private. Competition is also steeper for fewer grant funds available. The question remains: If the federal government is unable to respond to individual human needs, who will do so? Will the private sector again take up the cudgel? Can foundations, corporations, and individual donors make up for the gaps in federal funding brought about by the Reagan administration? What will the impact be on grants to individuals and small nonprofit organizations?

The New Federalism

The new federalism really means a smaller federalism or "let others step in."[6(p.38)] It started over a dozen years ago as an attempt to consolidate specific categorical grants into broader block grants in order to give states and localities more leeway in the spending of federal dollars. The current version has been dubbed "power to the statehouses."[7]

It has been estimated that under the new federalism there may be 37% fewer grant dollars available to individuals and nonprofit organizations.[7] All block grants contain provisions that federal funds supplement but not substitute for state and local funds. Another expected result is a great disparity of programs among states. If you need grant money and you live in a stingy state, you may wish to relocate.

What does all this mean to the individual grant seeker? The new federalism may be good for the grant seeker since it brings the funding process down to the local (hence, more controllable) level. Certainly fewer grant dollars will be available; competition, always quite stiff, will intensify. Federal officials hope that private philanthropy will step in to close the gap between programs previously funded by federal dollars and now abandoned to the control of states and localities. Indeed, initially at least, some grant seekers may experience new spurts of income as "would-be saviors" (private funders) come out of the woodwork.[6(p.34)] New and more creative grant ideas may evolve since sometimes the mere threat of cutbacks results in more inventive ways of doing things.

Those who believe that the federal government already has intruded too far into areas properly the domain of private philanthropy point out that it is really okay for certain projects to die a natural death if their originators can't drum up enough interest at the local level in order to continue receiving funding. The process of attrition may open the way for the funding of new and potentially more worthy grant ideas proposed by individual applicants. The field, at least in the beginning, could be wide open.

Other Trends in Government Funding

As mentioned earlier in this chapter, categorical grants, or money set aside for broad-based funding in various subject fields, are on the way out, both at the federal and the local level. These grant programs most likely will continue to decline during the eighties, to be replaced by project grants, formula grants, and government contracts.

Project grants are funds set aside for very specific short-term projects or the delivery of specialized services, e.g., an individual sociologist might receive a project grant to produce a fifty-page report on the impact of federal labor practices on the growth of local cottage industries. Project grants may include fellowships, scholarships, traineeships, and publication and research grants. A curious side effect of federal budget cutting is that such grants continue to remain popular. Even though the initial response of government funding executives was to hold firm on their grants to institutions while cutting back on their programs for individuals, it soon became apparent that there were strong arguments for doing precisely the reverse. Project grants have the advantage of being finite and their results may be highly visible within a relatively short period of time, thus providing positive public relations value for government funding executives. The recipients of project grants tend to view them as one-shot funding requests, whereas institutional applicants for categorical grants tend to come back year after year requesting additional funding. For these reasons, the trend toward project grants (and consequent support of individual applicants) should continue strong during the next several years.

Formula grants refer to funds for various educational, health, and welfare services returned to local communities under general revenue sharing or directly allocated to states according to specific distribution formulas. Formulas are based on such factors as how many people live in the state, what income they earn, and what taxes they pay. This type of block grant represents a strong trend in federal funding and an important element of the above-described new federalism. Individuals can benefit from formula grants only by applying at the local level through a sponsoring community organization.

Contracting is another current trend in government funding. Under this concept, if an individual or private agency can provide an equivalent or better service at a lower cost than the government now provides, that individual or agency may contract with the federal government to do so. A contract is defined by the federal government as an instrument to *procure* research or some other type of work, whereas a grant is defined as a mechanism to *support* such activity. Examples of the kinds of services an individual specialist might provide under a government contract are computer programs, feasibility studies, questionnaire-type surveys, evaluations of ongoing projects, and so on. In other words, it may be possible for you or your sponsoring agency to contract with the government to perform the same work for which you might otherwise have applied for a grant.

Under the mechanics of contracting, state officials provide formal contract approval whereas federal regional offices provide policy statements and guidelines. Usually, but not always, applying for government contracts involves highly competitive bidding. Contracts are advertised by requests for proposals (RFPs) announced in the *Commerce Business Daily* (see Chapter 6). Most RFPs appear in September, and a surprisingly short response time is permitted for the applicant.

Although, under certain conditions, individuals are eligible to receive government contracts, they require a substantial investment of time and energy. For the experienced fund seeker, and those with strong organizational backup, government contracts present a very attractive potential source of financial self-sufficiency. Some winners of government contracts point out that they differ from grants primarily on the issue of who initiates them. From the grantee's standpoint, nevertheless, grants are preferable to contracts, because they are not as competitive and more flexible in the use of the money, and there tends to be a lesser degree of monitoring performed by the granting agency.

STYLES OF APPROACH TO GOVERNMENT FUNDERS FOR THE INDIVIDUAL GRANT SEEKER

Government funding agencies differ from private funders insofar as the approach to them generally is more formalized, including step-by-step procedures you are required to follow. Because of the many procedures involved and the frequently complex reviewing process, government agencies are slow to act on your proposal. Particularly at the state or local level, it is often nearly impossible to determine in advance exactly when you will hear their decision.

Unlike private funders, government agencies almost always use some type of printed application form. Detailed instructions are usually provided and there is little room for free-form creative writing. When you first examine a government funding application form, nonetheless, you may be surprised to note how closely it corresponds to the components of the typical proposal (described in Chapter 7). Filling out application forms is a skill that tends to improve with practice and repetition. For this reason, and because, in some years, grant money is rotated within fields, it may be worthwhile to continue reapplying even after one or two rejections. It is truly a challenge for you, the grant seeker, to strike the proper delicate balance with your application form. You want to follow the rules and portray yourself as a legitimate, highly qualified, and serious applicant. At the same time, you wish to do something in the brief pages provided to make your proposal conspicuous among the hundreds or thousands of other proposals.

Attitude

Keep in mind when applying for government funds that you are requesting public dollars, expended under specific legislative appropriations and authorizations by laws, acts, bills, or resolutions. One reason for all the red tape is that legislative authority must by its very nature be closely guided by certain rules, regulations, and written procedures.

When approaching government funders at any level, you may find it expedient to adopt the attitude of entitlement: government funds are monies to which you have at least as rightful a claim as anyone else. Since you are entitled to this money, you simply need to find out the best way to go about getting it. Don't be afraid to ask questions. Request clarification of items you don't understand from the receptionist at an agency office or from staff members, the secretary to the director, or even from the agency head himself! Go on up the line until you receive a satisfactory answer to your question. Once you have determined that a government program exists under which individuals are eligible, the main obstacle standing in the way of your receiving funding should be the manner in which you apply. Persistency, aggressiveness, and proper understanding of procedures all contribute to improving your application.

Personal Contacts

For success with government funding agencies, it is important to make the right personal contacts. Contrary to the manner in which private funders operate, government grant officers tend to be more accessible to personal meetings than foundation or corporate executives and more willing to negotiate terms once they meet with you. Of course, the higher up your contact is, the better. However, beware of pitfalls. Government bureaucrats are easy to get to see. The hard part is determining who has responsibility for what and who the real decision makers are.

In order to help cut through the red tape, one approach is to adopt someone in each funding office as your "telephone buddy." This person need not be high in the government hierarchy; your telephone buddy may be a clerk or a secretary. All that matters is that he or she be able to explain the funding program and application procedures to you. Once you've submitted your application, try to maintain close ties with your telephone buddy. Although this will not actually help your chances of getting a grant, it will keep you informed as to the status of the decision-making process regarding your proposal and when you might expect to hear the results.

DECISION MAKERS

Finding out who's in charge is most difficult with government agencies. All too often the "pass the buck" mentality prevails, and no one seems willing to admit direct responsibility for the program in which you have an interest. Upon inquiring, you may be told that a particular program, though publicly announced, is not yet funded or that all funds have been expended. Of this you can be reasonably certain—whenever you apply to government funding agencies, complex political processes are at play, which you will never fully understand and over which you have little control. Try not to be intimidated by your own lack of clout or political influence over decision makers.

How to Tell Who's in Charge

Decision makers in federal government funding agencies often have titles like "Grant Administrator" or "Funding Program Officer." For federal programs the *Catalog of Federal Domestic Assistance* will tell you who is responsible in each case. It is also possible to call the Federal Information Center nearest you to find out. (See Appendix.) When in doubt, go to the top, that is to say, contact the office of the director of the specific agency under which your program falls. Refer to the most current *U.S. Government Manual* for assistance in identifying the proper agency.

For state and local agencies, the problem of identifying decision makers is complicated by jurisdictional disputes. It is helpful to use your own personal contacts, if any, to find out who's in charge. If this fails, again go to the top—contact the office of the mayor or the governor to find out who administers your program. Don't overlook the office of your local Congressman. Congressional staff can help direct you to the proper state or local official and sometimes even help with the actual application. In some cases, you might receive government funding simply because of the political leverage of your Congressman, because it is the only application submitted from your geographic locale, or because no funds were awarded to your region last year.

THE REVIEW PROCESS

Government agencies *always* employ a formal review process. Many permit a procedural walk through prior to formal submission of an application. Since procedures vary from agency to agency, you should ask about the review process in advance of submitting your proposal at the federal level. Such procedures usually include: (1) a staff level reading (or initial screening) of your application to see if you've fulfilled all requirements, (2) outside review by professional evaluators, subject specialists in your field, and/or a panel or jury of your peers, and (3) an internal funding conference during which the merits of competing applications are each briefly weighed. In most cases, the prior recommendations of the reviewers and/or evaluators are respected when the final decisions are made. At the state and local level, decisions are made by a committee or by the director of the agency. Although less formalized, the decision-making process at local funding agencies also relies heavily on the recommendations of experts or outside consultants.

Timing

Timing is an important element of the review process. Always submit your application by the deadline date, and well before the deadline, if possible. Unlike private funders, government agencies publicly announce deadline dates well in advance and usually will not accept an application that is even one day late. Once you've submitted your application, expect the review process itself to take up to three months. If you are successful, it may be even longer before you actually receive a check.

There is a phenomenon known as *fourth-quarter acceleration* that is of potential benefit to the individual grant applicant. If you fail to get your application in on time, or in some cases, even if your grant request has already been turned down, you may have a second chance during the month of September.

The fiscal year of most government funding agencies ends October 1. Thus, in the early fall there may be a rush within certain government agencies to expend funds that are still left over. (This can happen even in times of restricted federal budgets.) Federal funds not used by October 1 may have to be returned to the U.S. Treasury. Government funding executives fear that if this happens, the next year's Congress will reduce appropriations for their agencies.

Hence, it is said that some government bureaucrats keep a stack of proposals in their desk drawers. These proposals are for grant ideas that may be pet projects of the funding executives but that previously did not pass muster for one reason or another. Leftover dollars from the agency's budget frequently are used to fund such projects during the last quarter (early fall), since this is viewed by government funders as a favorable alternative to returning the money to the federal treasury. For the grant seeker, getting your proposal into this stack is a question of proper timing and establishing the right contacts.

THE FEDERAL GOVERNMENT FOUNDATIONS

The federal government has been called the "largest foundation of all." Although the words *government* and *foundation* may seem to conflict with one another as they've been used in our discussion of funders so far, this term truly describes the National Science Foundation (NSF) and the National Foundation for the Arts and Humanities, itself composed of the National Endowment for the Arts (NEA) and the National Endowment for the Humanities (NEH). These are federal government agencies that operate with public tax dollars but disburse grant funds to nonprofits and individuals much in the manner of the large private foundations.

The National Science Foundation

Established in 1950, the NSF promotes and advances scientific progress by sponsoring scientific research, encouraging and supporting scientific education, and fostering scientific information exchange. Under its province, NSF includes mathematics, physical, environmental, biological, social, behavioral, and engineering sciences. Outside its sphere of interests are clinical medicine, arts and humanities, business, and social work. Basic and applied research are funded, although at present, as in most government funding, applied research is favored.

Total grants awarded by NSF during 1982 amounted to approximately $100 million. These included scholarships, graduate fellowships, and traineeships as well as numerous grants of varying sizes for sponsored research at a wide range of institutions. The main drawback to

NSF grants is that, for the most part, unaffiliated individuals need not apply. Only a few of the 12,000 to 13,000 new grants awarded each year by NSF go for individual projects without specific institutional sponsorship. Exceptions are made for those unaffiliated scientists who apply for funding for "meritorious research" *if* they have the capability and access to facilities needed, agree to fiscal arrangements satisfactory to NSF grants officers, and otherwise meet the conditions set up by NSF policies. It is difficult to quarrel with the institutional affiliation requirements, moreover, since most scientific research requires very specific equipment, resources, and backup personnel, which only a qualified institutional sponsor can provide.

The balance of NSF grants go for the traditional mode of scientific research within the university. Most of the awards permit individuals or small groups of faculty to carry out projects in their own laboratories. The sponsoring arrangements favored by NSF, nonetheless, may be of benefit to the individual grant seeker in other ways. For example, NSF has given over $4 million in seed (start-up) money for experimental innovation centers at universities across the country. These centers, in turn, run programs to help inventors develop their ideas into marketable projects.

Like most federal agencies, NSF routinely uses outside consultants to review proposals. Preliminary letters of inquiry are recommended before you submit a formal proposal. They even have staff members who will sit down with you and help you to apply, if you are eligible and if your project falls under the province of the NSF. Contact the NSF directly in Washington, D.C., for further specifics on their programs. Ask for the current *Grants for Scientific Research* and *Guide to Programs.*

The National Foundation for the Arts and Humanities

In 1965, Congress created this government foundation that actually consists of two separate endowments, NEA and NEH. The two Endowments share a staff for administrative functions but have two separate budgets and independent program staff. Each has its own Council that advises on grant policies and programs. Both call upon panels of experts to help evaluate applications. Grants are awarded by a "jury of peers."

Since their formation, the Endowments have worked successfully to broaden and diversify the public's experience of the arts and the humanities. They have received much public scrutiny and attendant publicity despite the fact that the combined budgets of NEA and NEH equal approximately 0.002% of the total federal budget.

A 1980 Harris poll showed that defense and cultural programs were among the few things Americans still agreed deserved increased public support. Besides, it would take another congressional act to disband the Endowments. A special presidential task force, consisting of prominent leaders of the arts and business community, was set up during 1981 to study the Endowments. Their recommendations basically support the continuation of existing grant programs via the Endowment structure. It was suggested, however, that some modifications be made in the peer review system (such as preliminary screenings of certain applications) as an economy measure in order to reduce the heavy work load placed on the reviewers of grant applications.

The National Endowment for the Arts

In its current *Guide to Programs*, NEA states that it serves as a catalyst to increase opportunities for artists and to spur the involvement in the arts of private citizens, public and private organizations, states, and communities. NEA provides three major types of financial assistance: (1) nonmatching fellowships for artists of exceptional talent, (2) matching grants to nonprofit arts organizations, and (3) grants to state and regional arts agencies. It has the following program areas: dance, design, expansion arts (community programs), folk arts, international literature, media, museums, opera/musical theater, theater, and visual arts.

The total appropriations for 1982 for NEA were about $143 million. Appropriations for 1981 were $159 million; appropriations for 1983 will be only about $100 million. During fiscal year 1981 (latest available figures), NEA budgeted about $10.9 million for approximately 1,000 grants to individual artists. These fellowships ranged from $1,000 to $12,500. Such grant amounts to individuals are relatively small but are enough to tide the artists over for a period and to provide the time and the incentive for further work and experimentation. Current grants include fellowships to jazz musicians, choreographers, translators, composers/librettists, photographers, writers, and visual artists to publish books, advance their careers, and for investigations, explorations, and research. NEA has been called the "single most generous benefactor outside of purchasers of work, to the individual creative artist in the U.S."[3(p.59)]

In addition to its programs of direct assistance to individual artists, NEA's institutional grants may also be of benefit to individuals. For example, NEA dance programs serve to advance the careers of individual (relatively unknown) composers whose works must be commissioned by the choreographers as part of the stipulations of the original grant awards.

In times of federal budget cutting, one would assume that NEA grants to individuals would be more severely affected than those to institutions. Frank Hodsoll, chairman of NEA, recently was asked this question in a *New York Times* interview.[8] He responded that presently only about 6% of individual applicants are funded by NEA, whereas 50 to 75% of institutions that apply are funded. Yet individuals usually come to NEA only once, whereas institutions tend to come back again and again, with the same ones continuing to receive funding. To quote Mr. Hodsoll's tentative conclusion: "So one has to ask the question given those percentages, as to whether or not we have erred too much in the direction of institutions vs. individuals." It seems to me that this statement at least introduces the possibility of increased support of individuals by NEA during the years ahead. For further information, contact NEA in Washington, D.C. Ask for the current *Guide to Programs* and Annual Report.

The National Endowment for the Humanities

In its current *Program Announcement,* NEH outlines four specific goals: (1) to promote public understanding of the humanities, (2) to improve the quality of teaching in the humanities, (3) to strengthen the scholarly foundation of humanistic study and to support research, and (4) to nurture the future well-being of institutional and human resources that make possible the study of the humanities. NEH programs cover the following fields: language, linguistics, comparative religion, ethics, history, criticism, and theory of art, humanistic social sciences, and the relevance of humanities to current conditions of national life. NEH supports both individuals and institutions. As with NSF, even though not a requirement, most grant recipients have institutional affiliation. NEH does not support predoctoral fellowships, research or study in pursuit of academic degrees, or individual travel to professional meetings.

During fiscal year 1982, NEH appropriations were $130 million. For the purpose of comparison, during 1981 they were about $150 million; for 1983 they will be about $96 million. Although the other five NEH divisions make grants both to individuals and to organizations, the Division of Fellowships and Seminars distinguishes itself by focusing its awards primarily on individuals. This division has seven programs benefiting individuals of varying backgrounds. Each year, approximately 2,600 fellowships ranging in size from $2,500 for summer stipends to $22,000 for 12-month periods are awarded. The purpose of these fellowships is to provide uninterrupted time for humanistic study, research, and writing. Recipients are required to devote full time to their grant projects. Applications from unaffiliated individuals are accepted as well as those from grant seekers with institutional sponsorship. Payment is made directly to the individual recipients.

Those receiving NEH grants under the Division of Fellowships and Seminars include established scholars writing books, younger scholars early in their careers, interpretive writers both inside and outside academia, faculty members at research institutions and at colleges, and those in the nonteaching professions, such as doctors, lawyers, and journalists. Fellowships are awarded for independent research, personal study costs, with or without eventual publication in view, projects to enhance an applicant's teaching abilities, stipends, residencies, and summer stipends.

NEH has been criticized in recent years for placing less emphasis on the work of individual scholars, with the larger percentage of its grants going directly to institutions. Perhaps, as with NEA, federal budget cutting will result in a reconsideration of grants to individuals vis-à-vis those to institutions. Many of the NEH programs of institutional grants also are of benefit to individuals. For example, the Division of Research Programs supports a limited number of conferences, symposia, and work-shops to enable individual scholars to discuss and advance the current state of research on particular topics.

At present, the proposal review process at NEH includes four stages: (1) staff review, (2) review by panelists of outside specialists chosen for their expertise in given subject matters, (3) consideration by the National Council, composed of 26 members appointed by the President, who meet four times a year to advise NEH, and (4) final decision by the Endow-ment Chairman who generally follows the recommendations of the panel-ists and/or Council members. Although this review process may sound imposing, try not to be intimidated by it. Approximately one-third of those who submit applications to the NEH Division of Fellowships and Seminars are funded. Given today's competitive grants market, these odds are not all that bad. NEH staff members encourage the submission of preliminary applications and/or inquiries so that they can help pro-spective applicants determine their project's eligibility and the degree of competitiveness. If your grant idea is deemed inappropriate for NEH funding, staff members will refer you to other federal agencies. If you are turned down, they will provide you with reviewers' comments to help determine whether or not you should reapply.

Contact the information officers at NEA and NEH in Washington, D.C., for further details on their programs. Ask for program announce-ments, current annual reports, and application forms. Do this well in advance of deadline dates so that you have time to ask questions and to revise your application according to input from the Endowment staff. These staff members have the reputation for being quite helpful to the grant applicant who is persistent in requesting information and aid. Make use of this valuable resource.

FOUNDATIONS

DEFINITION OF A FOUNDATION

A foundation was once defined as "a body of money surrounded by people who want some." Although the originator of this definition is anonymous, it has been attributed to various foundation executives. The Foundation Center defines a foundation as "a non-governmental, non-profit organization with funds and program managed by its own trustees or directors and established to maintain or aid social, educational, charitable, religious or other activities serving the common welfare, primarily through the making of grants."[9(p.3)] The two elements of this definition important to you as a grant seeker are that organizations fitting this definition have their own funds (e.g., they are not engaged in fund seeking themselves) and that they actually make grants to individuals as well as to institutions.

This definition once noted, it must be acknowledged that foundations distinguish themselves by being as difficult to pin down and to make generalizations about as any category of funders. The term *foundation* covers a diverse group of organizations, ranging from a funder operated out of an accountant's desk drawer that exists primarily on paper as a tax benefit to the donor to a huge international institution like the Ford Foundation with its own internal bureaucracy.

TYPES OF FOUNDATIONS

The introduction to the current edition of *The Foundation Directory* distinguishes among four different types of foundations:

Independent foundations are funds or endowments designated by the IRS as private foundations whose primary function is the making of grants. They are often called *family* foundations because most derive from gifts of one individual or family and many function under the voluntary direction of family members.

Company-sponsored foundations are independent grant-making organizations even though they maintain close ties to the corporation providing the funds by means of endowment or by annual contributions. Officers of the company as well as other persons not affiliated with the company may serve on their boards. Company-sponsored foundations make grants on a broad basis. (This distinguishes them from corporate giving programs, which are more closely tied to the parent corporation, both in administration and in grant-making activities.)

Operating foundations are designated by the IRS as private foundations even though their primary purpose is to operate their own research, social welfare, or other programs as determined by their governing bodies. They sometimes award small grants to outside agencies or to individuals.

Community foundations, designated as public charities by the IRS, still operate much like private foundations. Their funds derive from many donors rather than from one single source, and grant programs are directed to their own immediate localities. Governing boards of community foundations usually are made up of representatives from various segments of the community.[10]

The terms *national, regional,* or *local* foundations refer to the scope of grant-making activities. Most foundations give only within their own locality or state. Only about one-third of the larger foundations, listed in *The Foundation Directory,* give on a national or regional basis.

The terms *general* or *special purpose* foundations reflect the type of giving or the program limitations of the foundation.

The terms *large* or *small foundations* reflect an "arbitrary, but convenient distinction which is based either on asset size or on total giving of the foundations."[11] To The Foundation Center, a large foundation has assets of $25 million or more and all other foundations are "small." The reason total giving is also considered is that some foundations are in what is called a *pass through* situation where substantial funds are not held in endowment. On paper, the foundation would appear to have little or no assets. Nevertheless, these foundations receive funds each year from their donors and expend most of this money in grants before the year is over.

Nearly half of all foundation grants come from the smaller foundations, although they account for less than one-third of all foundation assets. The typical small foundation grant averages anywhere from $3,000 to $11,000. The large foundations receive more proposals than the smaller ones and can afford to be highly selective. They prefer grant ideas that will function as models or prototype projects with national impact in particular fields. Smaller foundations are more likely to fund general operating budgets for longer periods of time.

As we shall see later on, from the point of view of the grant seeker, whether a foundation is staffed or unstaffed has more impact on the style of approach you adopt than its size or which category it falls under.

FACTS AND FIGURES

There are approximately 22,000 foundations in the United States, and most are quite small. The combined assets of all foundations exceeds $35 billion, and they currently expend about $2.6 billion in grants annually. By means of comparison, the federal government has been spending more than $100 billion each year on the arts, education, health, and welfare, although government grants may be sharply curtailed in the future (see the previous section). For the first time in 1979, corporations began to edge ahead of foundations in terms of annual giving. As will be discussed later on, corporations are the one category of funders that appear still to be on the rise.

In the current *Foundation Directory* (8th ed., 1981), there are 3,363 foundations with assets of $1 million or more and/or annual giving of at least $100,000. This represents an increase of 225 over the earlier edition of the *Directory* (7th ed., 1979); yet their assets are worth 30% less. In fact, with inflation and the state of the stock market, foundations have slowly been slipping in terms of granting power over the past few years. The Economic Recovery Act of 1981, however, offers some potential relief. This Act establishes a flat 5% (of market value assets) as the minimum amount foundations must award annually in the form of grants. The former percentage requirement was higher and perceived as punitive by some foundation managers. Foundations now have the option to distribute all their investment income annually or to reinvest all or part of their earnings in excess of the 5% figure to strengthen their future granting capacity and keep pace with inflation.

No one can predict precisely what role foundations will play in the mid- to late 1980s. Contrary to some dire predictions, there do not appear to be a huge number of dissolutions of foundations. Nor are there many instances of foundations turning over their assets to nonprofit institutions, like schools and hospitals. Most foundations will continue on in the business of giving money away. The small foundations will make a lot of small grants in their own geographic areas; the large foundations, although few in number, probably will continue to hold nearly 50% of all foundation assets and to award grants on a national and international basis according to highly competitive and specific criteria.

The important fact for you, the grant seeker, to keep in mind is that depending upon the nature of your grant idea (and its projected impact) you may be applying to a large national foundation or to a small local one, and different styles of approach (discussed later on in this section) may be required in each instance. In analyzing specific foundations as potential sources of grant money, their size and type are not as telling as their past patterns and their policies regarding grants to individuals.

How Good Are Your Chances?

Each year approximately 1 million proposals are submitted to the 22,000 foundations and, according to one informed source, only about 6 to 7% are actually funded.[9(p.xiii)] Some of the larger, staffed foundations report a turndown rate of 80 to 90%. Estimates of rejected proposals at unstaffed foundations run as high as 99%.

Do not be discouraged by these claims of astronomical rejection rates. They include all of the applicants who simply sent off mass mailings to foundations and all of the inadequately prepared proposals sent to inappropriate funders. If you have carefully selected certain foundations as potential funders of your grant idea, and if you have done your homework and prepared a solid proposal, my guess is that your chances are much better than you might think. This is partly where luck comes into play. However, you can engineer your own good luck. One of the purposes of this book is to help enable you to create funding opportunities where none seem to exist. You can lower the odds against you through research and careful selection as well as with well-documented proposals. If a foundation has money and has made grants to individuals in the past, or if foundation management appears willing to do so now, your job is to find a way to make it happen for you and for your grant idea.

The introduction to the current edition (1982) of *Foundation Grants to Individuals,* nevertheless, is really quite pessimistic. The editors predict that with federal government cutbacks the number of individuals seeking foundation support will increase. However, there does not seem to be a proportionate increase in the small number of foundations making grants to individuals. The expected result, therefore, will be increased competition among applicants. The editors note that getting a foundation grant is a "difficult task" because of the small number of foundations with limited assets, many limitations on their grants, and restrictive federal laws governing them. They predict that no substantial increase in grant amounts is likely to occur in the near future.

The lesson to be drawn from such gloomy predictions is to know your funder. The editors of *Foundation Grants to Individuals* advise that you make sure your grant idea falls precisely within the stated qualifications and requirements of each foundation you apply to. If you don't match up exactly with their restrictions, you'll probably receive a rejection or no response at all. They also recommend, as I have elsewhere in this book, that you use your proposal to demonstrate your familiartiy with the foundation's programs and how your grant idea fits in with that particular foundation's guidelines.

I feel that the tenor of this advice is a bit too negative. Although I agree that careful research is essential for the individual grant seeker, and I deplore the common practice of simply sending out proposals to foundations on a mailing list, I advise you not to take the guidelines and exclusion policies of foundations too literally if you can find a way for your grant idea to work around them. Foundations are notoriously individualistic in their funding styles. As compared to other categories of funders, they are relatively flexible in their ability to act on a proposal or to shift priorities if their managers so choose. So be creative. In your search for foundations as prospective funding partners, don't dwell on the negatives. Find a way to enable them to fund you. If this means affiliating with a sponsor, forming a coalition with other grant seekers, getting community support, or modifying your grant project in certain small ways, you should be willing and ready to do this. You need to be flexible along with the funder.

Although the pool of foundation funds available to individuals is not increasing, the editors of *Foundation Grants to Individuals* in an earlier edition noted an "equalization of access to the grant making process."[12] What this means, no doubt, is that the Tax Reform Act of 1969 and regulations published in 1976 and in subsequent years forced foundations to be more fair and open regarding their programs of grants to individuals. The Foundation Center maintains that more widespread information available on foundation grants will result in higher caliber and more suitable applications from grant seekers. This is my belief also and one of the desired results of this book.

THE TAX REFORM ACT AND FOUNDATION GRANTS TO INDIVIDUALS

The Tax Reform Act of 1969 and subsequent regulations provided the basis for the Internal Revenue Code Section 4945, which prohibits grants to individuals unless they are awarded on an objective and non-discriminatory basis under a procedure approved in advance by IRS. Grants must be (1) scholarships or fellowships; (2) prizes or achievement awards (but not contests or payments for expected future services); or (3) money to enable the recipient to achieve a specific objective (e.g., produce a report or other similar project) or to improve or enhance artistic, musical, scientific skills, or talents of the grantee or his teaching capacity.

Along with requiring foundations to submit their selection criteria for grants to individuals in advance for review and one-time approval, IRS also requires additional record keeping and follow-up on the part of foundation managers to ensure that the grant dollars are used as intended. This is the only instance where the IRS requires funders to submit grant policies in advance. It is done to ensure against abuses (e.g., the infamous case of a dentist who set up a foundation, appointed his wife as paid executive, donated his house to the tax-exempt organization, and awarded only three scholarships—one for each of his children) and to protect grant seekers from programs that are arbitrary or unfair.

Although the stipulations set up by the Tax Reform Act of 1969 and refined by subsequent IRS regulations at first seemed to be a matter of concern for grant seekers who feared such close IRS supervision would result in the cancellation of those few existing foundation programs for individuals, the years since 1969 have proved such fears to be unsubstantiated. The number of foundation grant programs for individuals seems more or less to have stabilized. The IRS regulations actually provide a form of insurance against unfair grant-making practices, which turns out to be genuinely beneficial for the grant seeker. Although the restrictions may have inhibited a few foundations from making grants to individuals, most foundations easily secured approval for their procedures and very few had to change the manner in which they already were awarding such grants.

WHAT IS THE RECORD ON GRANTS TO INDIVIDUALS?

Over 44,000 individuals annually receive grants from foundations. (Over 50% of these grants are broadly defined as for "educational purposes.") Here are some examples of foundation grants to individuals, chosen to indicate the wide spectrum of such grant programs.

• John Irving wrote *The World According to Garp* and E. L. Doctorow wrote *Ragtime* on money derived from John Simon Guggenheim Memorial Foundation grants.

• The Wonder Woman Foundation, founded by Warner Communications and DC Comics, provided 20–60 awards during 1982 to women who represent the qualities and characteristics of the fictional heroine Wonder Woman, i.e., they are peace loving, independent, honest, courageous, compassionate, and wise. In addition, grants were given to women over 40 years of age, based on need and "realizable promise." The use of the grant funds is unrestricted (but presumably most of the money will go for education, career development, or projects furthering the cause of equality for women).

• The Debbie Fox Foundation of Chattanooga, Tennessee, provides assistance, based on medical expenses, need, and other criteria, to individuals with severe facial deformities.

• The Rockefeller Foundation provides grants of $9,000 each to eight playwrights for six-week residencies at the theaters of their choice.

• The Bush Foundation of St. Paul, Minnesota, has a program of leadership fellowships for mid-career development awarded to persons between the ages of 28 and 50 who appear likely to advance to top positions in the fields of business, engineering, forestry, government, journalism, law, theology, or administration in arts, education, or health and who are residents of Minnesota, North Dakota, South Dakota, or parts of Wisconsin. These stipends of $2,000 per month plus 50% payment of tuition and fees may be used for further training in the aforementioned fields, particularly in the area of administration

• The Money for Women Fund of St. Augustine, Florida, in a recent year awarded six grants ranging in size from $250 to $1,200 to individuals in financial need to achieve greater self-development through the arts or to improve or enhance a literary, artistic, music, scientific, or teaching talent or skill.

• The Frank Gannett Newspaper Carrier Scholarships Foundation of Rochester, New York, awards over 400 grants annually, averaging around $4,000 each for a 4-year college course (in payments of $500 per semester) to individual recipients who were carriers of Gannett newspapers for at least one year preceding the date of the award. (The moral is—if you're a high school student and you need a scholarship, get a paper route for a Gannett newspaper!)

• The S & H Foundation of New York has a Merchant Scholarship Program under which grants are awarded to 500 high school seniors whose parents own retail businesses that give out S & H green stamps or who work for merchants who give out the stamps.

As you can see from the above more or less random sampling of foundation grants to individuals, they are characterized by nothing short of complete diversity. They are also quite specific in intent, with many restrictions and limitations. (As mentioned elsewhere in this book—the grant seeker should *always* find out about these in advance of making application to any funder.) At the same time, foundations tend to have a high degree of flexibility and the ability to act if their managers perceive a need to be met or if a particular project appears worthwhile. This is in sharp contrast to government and corporate grant programs, which are tied down by bureaucratic procedures and profit-related limitations, respectively.

What Kinds of Grants Do Foundations Make to Individuals?

As the current (1982) edition of *Foundation Grants to Individuals* indicates, most foundation dollars for individuals go for education of one form or another. These awards are highly specialized according to either subject discipline or grantee population. Approximately 1,000 foundations are included in this edition and most have very specific limitations.

By far the largest category of individual grants is for *scholarships* for undergraduate study, and most of these have geographic require-ments (i.e., scholarships are given only to residents of certain states or localities). The next largest category is composed of *fellowships* or grants for independent projects or research as well as for work related to doc-toral dissertations. These are distinguished by their specificity as to sub-ject discipline rather than geography. The next largest category is *grants restricted to company employees and to recipients who reside in particu-lar localities.* These include corporate foundation grants to employees, former employees, and their families. Such grants usually provide schol-arships or pensions and aid in medical or living expenses rather than money for creative work or research projects. To qualify for these grants, family relationships need not necessarily be close, so remember to shake your family tree for a corporate connection. The next two categories of foundation giving to individuals are roughly equal in size and numbers of grants. They are (1) *grants for general welfare and medical assistance* for emergency or long-term personal, medical, or living expenses (many paid through hospitals or social agencies) to individuals with particular professional or ethnic backgrounds or facing special kinds of hardships and (2) *awards, prizes, and grants through nomination,* grants usually in recognition of prior achievement for which institutions must apply on behalf of individuals. (Remember it is possible to seek a nomination or to gain such institutional support, as indicated in Chapter 2 of this book.) A final category is *internships and residencies,* which provide studio space, research or conference facilities, equipment, materials, or other cash equivalents to enable individuals to pursue creative or scholarly endeavors. Such programs may also include grant dollars as supplemen-tal stipends or salaries. Most foundation grants to individuals are for U.S. citizens. As the third edition of *Foundation Grants to Individuals* indi-cates, there were only eight grants given to foreign individuals.

THE RELUCTANCE OF FOUNDATIONS TO SUPPORT INDIVIDUALS

Tradition has it that foundation managers are loath to make indi-vidual grants, and there are several reasons why they hesitate to make grants directly to individuals. The fact is that, at least up until now, foundations have received better written, more concise, and to-the-point, that is to say, more "professional," proposals from organizations rather than from individuals. As we will see in Chapter 7, this does not have to be the case.

Foundations also hesitate to give to individuals because, by making a grant to an institution instead, their managers dilute their responsibility in case of dissention with the IRS. If the dean of a college runs off with the building fund for the new library, the foundation executives responsible for donating the money will not be held as immediately accountable as if the recipient of a research or fellowship grant absconds with the money instead to South America. Foundations are obliged by IRS to take steps to recover grant funds if misused. As a penalty for failure to comply with IRS regulations, in some cases, excise taxes may be imposed on foundation managers themselves.

As you can see, at least from a foundation executive's standpoint, awarding grants to institutions is safer and simpler than giving away money to individuals. Some foundation managers view individual grant recipients as simply too much trouble. For those foundations without full-time staff (and most fall in this category), there is much work involved in the processing of unsolicited proposals from individual applicants. IRS regulations require far more careful supervision and more paperwork (reports, bookkeeping, inspections, etc.) on the part of the foundations for programs of grants to individuals than for their organizational counterparts. Foundation executives must guarantee that they will follow up on their recipients to see that grant money is used as intended, and they must conduct a constant monitoring of their grantees' progress. All of this is much easier if the recipient happens to be a nonprofit organization with an established record-keeping apparatus as opposed to an individual artist, inventor, or "absent-minded" professor reputedly bad at keeping deadlines, adhering to budgets, and making reports.

TRENDS IN FOUNDATION FUNDING

We have already noted that foundation grants are characterized by diversity and limited geographic focus. Most foundations prefer to fund a specific project (rather than a general operating budget), which corresponds to a demonstrated need (as opposed to a brand new idea that has not yet been proven), which either terminates or will be self-supporting in some way at the end of the grant. Foundations prefer outright awards as opposed to loans, matching gifts, or other forms of reimbursement. Most grants are of one year's duration although repeat funding is a typical pattern.

Some foundations specialize in demonstration grants, others in research or publication. Some will fund your living expenses for the duration of the grant project, whereas others will cover only certain aspects of your budget. Some require that you have an institutional sponsor; others make out the check directly to the individual. The best way to determine what types of grants a particular foundation makes is to do your homework. If the foundation itself does not state its intentions in print, a careful perusal of several years' lists of grants for that foundation (see 990 forms discussed in Chapter 6) should provide an adequate picture of the types of grants awarded.

Foundations as a group have had their share of critics over the years. In addition to charges of tax dodging and unfair grant-making practices, which a number of the regulations set up by the Tax Reform Act of 1969 were intended to correct, foundation managers have been called elitist and insulated. They have been accused by critics on the left and the right both of concentrating on conventional recipients and of championing radical causes. There is some truth to the accusation that foundation managers tend to shy away from the controversial or politically charged grant idea. For this reason, their grants have been referred to as "band-aids" or small amounts of money, which temporarily stop the problem from hurting but don't really change the basic situation.

Foundations also prefer to step in and out of grant projects with start-up or seed money rather than to commit themselves to a long-term expenditure. They continue to be moving away from participation in large-scale projects and toward more modest ideas that will benefit specific population groups. This trend and the current emphasis on project grants limited to specific subject fields with priorities in the areas of higher education and training cannot help but benefit the individual grant applicant whose grant ideas tend to qualify as relatively small-scale "easy in/easy out" projects. Foundation managers hope that their grants to individuals and small groups will have a "multiplier effect," encouraging others to donate as well and enhancing the fund-raising capabilities of the recipient.

At recent conferences, foundation executives have reflected an underlying anxiety and confusion about their future roles. Some observers of the philanthropic community believe that foundation giving will decline in the years ahead.[13] Others predict a decrease in the numbers of foundation grants but an increase in the size of each grant. One trend is clearly indicated—a marked increase in competition for those foundation grants available to individuals and other applicants. You can be sure that grants decision makers will look even harder at each new proposal.

Foundation managers complain of an unrealistic expectation (at least according to them) on the part of the public and government officials that foundations (and corporations) will somehow find the funds to make up for federal government cutbacks in a wide range of programs. As a result, another trend that is developing is the formation of partnerships. Foundations will look for "local partners," such as nonprofit groups, corporations, and community agencies, to coalesce forces (in terms of both matching dollars and staff hours) in order to increase potential grant-making power. For the individual grant applicant, the implications of this trend will be the increased necessity of finding sponsoring organizations and providing strong evidence of community support for grant ideas at the time of application.

STYLES OF APPROACH AND GRANTS DECISION MAKERS

Bearing in mind all of the contradictory advice you will receive about approaching foundations, how should you proceed? Earlier in this section we noted that depending upon the size or the type of the foundation to which you are applying you may want to adopt a different style of approach. This is particularly true of the differences between staffed and unstaffed foundations. Whereas the majority of foundations have no paid professional staff, the small number that do have staff tend to control a goodly percentage of foundation money and include most of the better-known foundations whose names you've undoubtedly heard. The moral is: first and foremost, find out if the foundation to which you are applying is operated by a professional staff or whether it is run by a volunteer board of trustees.

Unstaffed Foundations

Unstaffed foundations, representing the vast majority of all foundations, tend to be more flexible in the distribution of small grants for individual projects. At the same time, their grants decision makers (usually nonreimbursed volunteers) are more susceptible to personal influence, e.g., if the head of the foundation already knows you, you're halfway there. With unstaffed foundations you should expect to submit a full proposal right away, since there is no staff member to speak with on the telephone and no one to respond to an initial letter of inquiry. Be prepared to wait (it may be months before you are notified), expect few, if any, chances to meet with foundation representatives, and, above all, learn the dates that the board meets. Since the board may meet only once or twice a year, it is essential that you get your proposal in well in advance of that date.

Grants decision makers may be members of the donor's family, friends, accountants, lawyers, bankers, corporate executives, or community leaders. They may have the titles "Directors," "Trustees," or "Foundation Managers." Depending upon their relationship to the original donors and the manner in which the foundation board is constituted, they may have more or less say in how foundation dollars actually are spent. Most foundations have officers and a board of trustees, but many are run essentially by one individual with a board that meets once or twice a year to rubber stamp his grants decisions.

The foregoing information might come as quite a shock to you. Up until now the word *foundation* probably conjured up the image of an institution in your mind. Most grant seekers are unaware that the majority of foundations are not managed in a manner that one could call "professional." There is no real foundation office, no receptionist to answer the phone, or secretary to respond to your request. More likely than not, the foundation is run out of the desk drawer of the aforementioned family friend or relation, who happens also to be an attorney or an accountant.

Staffed Foundations

Staffed foundations tend to prefer an initial letter of inquiry regarding their potential interest in your grant idea (see Chapter 7). They respond fairly quickly and generally interview major contenders for grants. At the time of the interview, some negotiation is possible, enabling you to modify your proposal to meet their requirements. As you might expect, grants awarded by staffed foundations tend to be larger and national, or at least regional, in expected impact. In general, for most grant ideas, the unaffiliated grant seeker would do well to stick with local foundations. However, if you do apply to a large, national foundation for a grant, be sure to stress the innovative aspects of your idea and the transferability of the results. How will your grant project serve as a model for others in different locations?

The approach to staffed foundations is far more formalized than that for unstaffed foundations, even though the eventual response time will be shorter. You should familiarize yourself with procedures at the staffed as well as the unstaffed foundations, because the former, although few in number, tend to represent the larger, more influential foundations, which make more than their share of grants to individuals, both in terms of numbers and dollars.

Most staffed foundations are relatively large. With some notable exceptions, they are located in the Northeast, many in New York. Like the unstaffed foundations, they too ordinarily have officers and a board of directors or trustees. Some of these foundations have well-informed boards who take an active part in the grant-making process, whereas others simply serve to validate the decisions of the better-informed professional staff, whose opinions are rarely questioned.

Keep in mind that not only is it bad form but it is also counterproductive from a practical standpoint to try to circumvent paid decision makers of a staffed foundation and to use your influence directly with board members you may know. When contacting staffed foundations, it is always best to "go through channels."

In a survey conducted by The Foundation Center of 3,400 large foundations, only 626 had staff and only about one-half of these were full-time. Most chief executives are white, male, and middle-aged, although the position of "program officer" appears to be a professional entry point for both minorities and women. One foundation president referred to the typical foundation official as "arrogant, aloof, lazy, and unwilling to venture into new areas."[14] Although this statement may seem overly harsh, you should be aware that the world of the foundation executive is a rarefied one. They tend to be somewhat insulated and self-protective about casual conversations with grant seekers, because everywhere they go it seems that someone wants something from them.

Making Application

Unlike government funders, most foundations do not use standardized application forms. A written proposal (designed according to the recommendations you'll find in Chapter 7) accompanied by a cover letter ordinarily suffices. A small number of foundations do require you to fill in application forms for their programs of grants to individuals, and an even smaller number prefer a personal meeting before you apply. Find out early on in the grant-seeking process which form of initial application the foundation prefers and proceed accordingly. If no instructions are available in this regard (most staffed foundations will send you their instructions on how to apply), assume that a written proposal is appropriate and submit it as soon as possible. However, keep it simple.

THE REVIEW PROCESS

Foundation executives report an alarming (to them) increase in the number of grant requests in recent years. Some estimate an increase of from 30 to 50% in the numbers of applications received. Foundations require a lead time of from two to six months (even more for some unstaffed foundations) to consider requests for funding. So select your foundations carefully, and try not to put all your eggs in one basket. As mentioned elsewhere in this book, choosing foundations as potential funders requires systematic and thorough research with a view toward discovering an extremely close match between your grant idea and the goals and intent plus the recent grant history of the foundations (particularly as to specific programs of grants to individuals).

Who Actually Reviews Your Proposal at a Foundation?

At an unstaffed foundation, the chief executive ordinarily decides whether or not to seriously consider your proposal. However, an initial screening more than likely will be performed by his secretary or another assistant, which is why your proposal and cover letter must be so carefully constructed. If your proposal passes this first test, copies of it are distributed to members of the board prior to their annual meeting, with or without the chief executive's comments.

At a staffed foundation, reviewers may be outside consultants, staff members, program officers, or a combination of the above. They may be your peers and subject specialists in your field, or they may know very little about it. Usually grants decision makers have some ties to the community in which the foundation is located and some experience reviewing proposals. They have all kinds of titles ranging from "Program Officer" to "Vice-President for the Humanities." Foundations are notorious for assigning ambiguous titles to staff members, such as "Executive Associate." Since most staffed foundations have listed telephone numbers and printed literature as well, the best way to determine who at the foundation will be the prime decision maker regarding your proposal is to inquire.

Regardless of who makes the final decision about your request for a grant, there are particular questions you may expect foundation proposal reviewers to ask. First and foremost, they will want to know how your proposed idea helps advance the goals of the foundation. Second, they will examine the merits of your idea as it relates to what others in your field currently are doing and what the projected benefits might be to society at large. Third, they will want to consider you as an individual. Are you responsible, highly recommended, and suitably qualified to carry out the work you propose? Finally, they will examine your proposal on its own merits. Have you demonstrated a need for the service you propose to perform? Are your objectives concrete and measurable? Is there a sound way to evaluate your performance? Finally, is your budget realistic in terms of what you hope to accomplish? (See Chapter 7 for further details.)

CORPORATE FUNDERS

FACTS AND FIGURES

Corporations use three basic mechanisms for giving away money: tax-deductible contributions, grants from company-sponsored foundations, and items treated as standard business expenses. By far the best opportunities for the individual seeking corporate support lie in finding a way for the company to treat your idea as a business expense item, although as we will see later on, individuals with *close family, geographic, or other ties* to corporations do get their share of funding through scholarships and occasional research grants (*if* the idea is perceived by management as being in the best self-interest of the corporation).

U.S. corporations traditionally have been permitted a charitable deduction of up to 5% of their pre-tax net income. The Economic Recovery Act of 1981 provides for an increase up to 10% for charitable deductions. However, most corporate donors still have a long way to go to reach even the 5% level. Only about 28% of the nation's 2.5 million corporations made any charitable contributions at all in the latest year of record. Most of those that did contribute donated only about 1% of their income.

On the other hand, corporations as a group have been steadily gaining in influence and importance when compared with other categories of funders. In 1979, for the first time corporate funders overtook private foundations in total grant awards. This trend appears to be continuing. Corporate foundations thus far have escaped the decline in assets that has plagued other private foundations. During 1981, corporations gave away $3 billion, an increase of about 11.1% over the previous year. About 60% of the 1,300 top industrial concerns (traditionally the major providers of corporate philanthropy) surveyed that year by the Conference Board reported that they planned further increases in their contributions for 1982 as well. This is one reason that savvy fund seekers have been turning their attentions more and more toward corporate funders.

It has been estimated, furthermore, that if corporations were to give at or near the level permitted by IRS, another several billion dollars would be made available for charitable causes.[15] Unless you have a qualifying nonprofit sponsor, however, the question of charitable contributions from corporations is a moot issue. Outright corporate gifts to unrelated individuals are few and far between, since corporations receive no tax deduction for such gifts.

WHERE DO CORPORATIONS PUT THEIR GRANT MONEY?

The answer is easy. Most limit their financial support to communities where they have business operations, either plants or headquarter offices. In a recent survey conducted among the nation's largest companies, 88% indicated a definite interest in services to youth, 78% in supporting social services to minorities, and 70% in hospital building programs. Arts and cultural activities were cited by 66.5%, hospital medical programs, 61.8%; health research, 61.5%; and employment training, 56.8%.[16] Surveys of corporate giving patterns can be somewhat misleading and should be viewed with some skepticism by the grant seeker. As with private foundations, there may be a discrepancy between what the funder *says* it is interested in and what grants actually are awarded.

Corporate grants tend to be small, the average ranging from $1,000 to $5,000; grants of about $500 are more typical for individual recipients. Corporate philanthropy has never been very adventurous. Usually it consists of last year's budget plus a few "safe" items thrown in. Managers of public corporations steer clear of controversial recipients, because they have to answer to stockholders, many of whom resist the concept of corporate philanthropy. For the most part, stockholders would prefer higher dividends, enabling them to make their own contributions to charity by their own choice.

There are advantages to corporate grants that are frequently cited by recipients. Among these are flexibility in utilizing the money and dependability and promptness of payment, as well as the fact that one corporate grant usually leads to another, either from the same or from a different corporation. Although corporate grants tend to be smaller than foundation or government grants, they also have fewer restrictions on the funds, especially regarding the use of grant money for general operating expenses as opposed to specific project-related expenses, e.g., you can use the grant money to pay a salary, hire a consultant, or contract for computer time. With other types of funders you (or your sponsor) might be limited to using grant money only for the direct costs of accomplishing your idea, e.g., for conducting an experiment, collecting data, or providing the agreed upon services.

WHY DO CORPORATIONS GIVE?

To quote Tom Boyd, a former corporate giving officer, now a consultant to nonprofit groups seeking funding: "Companies are private and can give their money away to whomever they please, whenever they want to, and in the amounts that suit them—or they can choose not to give any money away at all."[17]

Corporations give away money for any one (or a combination) of the following reasons: to get a tax write-off, to create a positive public image, to influence legislative or other opinion makers, to cultivate stockholder's good will, to build community relations, to please special interest groups, to return favors, to keep up with the competition, to perpetuate the past, to support employee services and/or training, to increase productivity, to entice prospective new employees or transferees, to ensure against future losses, and to associate with quality.[18]

Getting a corporation to give you money requires creativity, persistence, and much ingenuity. First, consider the motivation of the corporate funder. What does management stand to gain by supporting you and your grant idea? How is your idea potentially profitable to the corporation? How will it serve their employees? How will your grant idea help the corporation accomplish some of the benefits identified in the previous paragraph? Emphasize these aspects in your proposal and in any contacts you may have with corporate funding executives.

WHAT IS THE RECORD ON CORPORATION GRANTS TO INDIVIDUALS?

Although corporations do have substantial grant programs in support of individuals, most are limited to employees, former employees, and their families and/or residents of particular geographic locales where the company has business operations. This assistance ordinarily is in the form of scholarships, pensions, or aid with medical or other living expenses. Corporate giving to individuals tends to be "routine" rather than creative, with many small grants awarded to noncontroversial applicants. The one exception to this rule is corporate funding of research and development projects by individual scientists or inventors, working on their own to develop new products, which might eventually be taken over by the company. However, these funds often are carried by corporations under administrative, as opposed to philanthropic budgets.

Corporate grantsmanship for individual applicants is further complicated by the difficulty in getting information on the programs for which you may be eligible. Although one would assume that a prime reason for giving away money would be to improve the corporate image in the public's eye, oddly enough corporations have been extremely secretive about their activities in this regard. It is nearly impossible to get leaflets or other descriptive literature telling you how to apply, or indeed, anything in writing from most corporations about their contributions programs. Rare is the corporate annual report that mentions giving programs in any but vague terms. One writer acknowledges reading over 200 corporate reports and finding only one that provided a specific dollar amount for contributions.[19]

If an individual has a direct family connection with a corporation and he needs a small amount of money to continue his education or for some other immediate problem, chances are he will succeed. So before you give up on corporate grants entirely, shake your family tree. Find out if a member of your family is employed or formerly was employed by a company that has a contributions program for which you are eligible.

SINCE CORPORATE GRANTS TEND TO BE SMALL AND HARD TO GET, WHY BOTHER WITH THEM?

There are a number of different answers to this question. Some are simple; others are more complex. First, once corporations become involved in grant making, they tend to become highly competitive. Their managers are extremely susceptible to peer pressure and most interested in what other, similar companies may be doing. Corporations like to travel in numbers, and one corporate grant almost invariably leads to another from a company in the same or a related industry. So if you can get just one corporate funding executive to part with an initial small sum of money, you've greatly increased your chances of funding your idea. Establishing a good reputation as a corporate grant recipient is similar to acquiring a good credit rating. It enhances your resume, and you can rely on it to help generate future support for years to come.

Second, corporations may well represent the last frontier in grant making. As noted earlier, corporate funders have been experiencing a strong spurt of growth and activity in recent years, at a time when federal funding is in doubt and foundations have suffered a decline in assets and in the purchasing power of their grant monies. The wise grant seeker will at least inform himself about corporate giving programs and procedures so that he is well armed for the future. (See Chapter 6 for a discussion of some of the tools and techniques required to perform the difficult detective work involved in getting information on corporate funders.)

Finally and perhaps most relevant to the individual grant seeker is the fact that outright corporate giving is only part of the story of corporate largess. There are many ways other than via grants and contributions in which corporations support individuals and groups. It has been estimated that above and beyond corporate gifts to worthy causes, anywhere from $600 million to $2–3 billion is spent annually on similar programs but deducted by corporations as business expenses. And "in-kind" assistance, defined as employees' services and the use of corporate facilities, amounts to at least an additional $1 billion annually.

Since corporate decision makers hesitate to make outright grants to unrelated individuals, the grant seeker must find ways to work around the restrictions to gain other forms of corporate support, most of which fall under the advertising, public relations, or general administrative budgets of the corporations, rather than strictly under the province of the philanthropic arms. The rest of this section will direct itself to these other types of corporate funding. We will refer, henceforth, to corporate "support" rather than to grants, since these other categories of funding do not strictly fit our definition of a grant.

How to Get a Corporate Funder to Treat Your Idea as a Business Expense

From the point of view of the grant seeker, whether the support you receive is in the form of a charitable contribution to your sponsor, a grant from a company foundation, or an item treated by the corporation as a business expense, the dollars at your disposal are exactly the same. So if the corporate funder you have in mind has a policy of "no grants to individuals," you may wish to find another way to get the corporation to support you, most commonly via the business expense mechanism. Corporations list all kinds of items as legitimate business expenses, so why not your grant idea?

• If you're a painter, sculptor, or photographer, many corporations commission works by local aritsts to decorate their offices and lobbies. In fact, there is stiff competition among corporations in some areas to see which has the best (or most) art on display. One tip is to introduce yourself to commercial interior designers in your locale so that they will keep you and your work in mind for their corporate clients who wish to purchase and/or display art work.

• If you are an inventor or researcher, and your idea potentially might turn a profit somewhere down the line, you may get a local corporation to treat it as a business investment rather than as a grant. This is particularly true if you are developing a product or service that would otherwise provide some competition for the corporation's own products and services. Be sure to speak with an attorney or patent specialist before you sign an agreement with a corporation. (More information on getting funders to treat you as an investment is provided in Chapter 8.)

• If you are a musician, dancer, mime, or other type of performer, you may get a corporation to provide support in the form of funds for entertainment for their employees. Some corporations have occasional noontime concerts as fringe benefits for their employees.

• If your grant idea is one that will have high visibility as a project benefiting the community (e.g., a local history project, radio or television production, a fair or festival), corporations might support you but handle it as funds for public relations. If you can get just one corporation to do this, others will follow suit, and the idea should snowball.

• If your idea is one that clearly benefits corporate employees (for example, physical fitness and drug and alcohol rehabilitation projects are quite popular right now), you may get a corporation to hire you as a consultant and treat your idea as an administrative expense to improve employee productivity.

• If your idea involves a printed publication, pamphlet, or program, you may get corporate support in the form of advertisements.

The above are some examples of the varied types of corporate support individuals may receive. They are often overlooked by those who criticize corporations for not giving away enough money. As you can see, although these funds are used by the recipients in much the same way as grant money is used, they are carried on company books as advertising, public relations, or personnel-related expenses.

Social Investments

Another type of support that corporations sometimes provide is called a *social investment*. (Foundations call them *program-related investments*.) These are investments that would not otherwise be made under customary lending standards or those in which social considerations play a substantial part in the investment decision. Since they are not grants, they are not subject to the same competition or vigorous application procedures, and tax-exempt agencies are not shown preference. Since they are not loans, they do not have to be paid back.

Examples of the kinds of social investments corporations might make are new businesses started by minorities or women, cancer research, drug projects, health studies or services, family planning or employment counseling, ecological or environmental projects, new inventions, product safety studies, and recreational services for local youths. Finding a way to translate your idea into one of these social investments is a true challenge for the grant seeker and/or his sponsor.

Since corporations ordinarily do not stand to make a profit on social investments, they must have some other reason for investing in you. That reason is often tied in with the modern concept of corporate social responsibility. The idea behind this concept is that corporations have both a moral obligation and a compelling need to deal with social problems in the communities where they have plants. In fact, corporate social responsibility is just another name for enlightened self-interest. The contemporary corporation really has no choice but to at least appear to be socially concerned, since, like any institution, the corporation is dependent on the good will of society for its continued existence. If you as a grant seeker can find a way to get a corporation to invest in your idea while at the same time receiving recognition for supporting something that is good for society, you can capitalize on this trend.

Gifts "In Kind"

Corporate contributions have been described as the giving away of dollars, time, and service.[20] A gift of time, service, or something else other than money is referred to as a gift "in kind." In the recent past, corporations have been permitting their employees (especially those in the middle management category) to take reimbursed time off from their regular jobs to work on community or other projects not directly related to corporate activities. If you have a qualifying sponsor, you can get help carrying out your grant project through these "released time" programs. Corporate employees will provide you with free expert advice on fiscal management, bookkeeping, accounting, product development, and public relations if you need it. The corporation benefits from such programs because they serve more or less as mini-sabbaticals for employees who then return to work refreshed by their outside activities and with improved feelings of self-worth.

In addition to time, in-kind gifts might include the use of equipment, facilities, and a wide range of services from the corporation to the individual. An example of such in-kind support is the Contemporary Music Project. A Massachusetts corporation records the works of young composers and makes these recordings available at no charge to radio stations throughout the country. This service costs the corporation little in terms of dollars, but it carries a high public recognition value for the corporation among the listening audience. At the same time, it has boosted the careers of young composers who are unable themselves to pay the high costs of recording or to attract the major record companies. Without this program, most of these composers would have little prospect of ever having their work recorded, much less broadcast.

Additional examples of support other than cash that corporations donate include free legal, accounting, or printing services, air transportation, building materials, office supplies, furniture and equipment, use of computers, electrical work, carpeting, and even food. As with social investments, you have a better chance of receiving this type of corporate support if you have a qualifying sponsor.

APPROACHING CORPORATIONS

A Success Story

Let's consider the example of one grant applicant who has thus far been highly successful in getting support from corporations for his idea.

Harry D. Williams II is a semiretired Connecticut home builder, whose wife became seriously ill and nearly died of bladder cancer several years ago. Mr. Williams hurriedly began doing research on what causes cancer and discovered that the basic mystery about cancer—how a normal cell changes into a malignant one—is still to be solved. Although millions of dollars are spent on cancer-related research each year, most is for developing treatments, with very little going to what Mr. Williams calls "carcinogenesis." He decided to get to work on his idea, but first he needed funding. When he learned that the American Cancer Society had been unsuccessful in tapping corporate philanthropy, Mr. Williams decided to go first after this group of funders. From what he knew about corporations, Mr. Williams began seeking a nonaltruistic motive for corporations to support him.

The first thing Mr. Williams did was to haunt the nation's medical research establishment, getting introductions from one scientist to another, until he had made the acquaintances of six highly regarded scientists who agreed to serve as his board of advisers. He then set up a fledgling nonprofit organization, which will pass the research funds along to the individual researchers, who will conduct the various experiments. These researchers may receive as much as $100,000 a year for five years.

With great audacity, Mr. Williams then began contacting corporate officers, talking his way into the executive suites of some of the giants in the chemical and oil industries. He then made his pitch: these companies have a profit motive in supporting his research, since until the mystery is solved, industry will lose millions of dollars a year on products that turn out to cause cancer in workers and/or consumers. So far Mr. Williams has a total commitment of $2.5 million from corporations. Some of this money came out of research funds and departmental budgets rather than from funds set aside for charitable donations. One corporate executive actually gave Mr. Williams $10,000 out of petty cash funds. "I don't believe in corporate philanthropy," says he.[21]

What can the grant seeker learn from Mr. Williams' example? Although not an executive himself, Mr. Williams had a pretty good idea how corporate managers operate. He took a highly personalized approach, aimed at the profit motive of specific companies in industries most affected by cancer. However, before he approached his potential funders, Mr. Williams did a lot of research and leg work. He informed himself about the current state of cancer research and, in particular, the funding of that research. He got several prestigious scientists to help him. He even formed his own tax-exempt organization. He then made his pitch based on a nonaltruistic approach. Once one corporate executive decided to support him, others joined the bandwagon. It is noteworthy that a substantial part of the funding he received came not from corporate contributions budgets but from research and development funds, administrative, departmental budgets, and even petty cash.

Now you may not have Mr. Williams' commitment, audacity, or time to spend on fund seeking. Your idea may not warrant the formation of its own tax-exempt organization with a six-man board of advisers. However, you probably don't need anywhere near $2.5 million either. My point is that you can adopt Mr. Williams' general approach—becoming well informed in advance, finding a profit-related motive for the corporations to support you, making a personal appeal, if possible—that will lead to success in the funding of your own grant idea.

The Importance of Personal Contact

With corporations perhaps more than with any other category of funder it does help to know someone, especially "someone on the board." Although the idea of a closely knit club of corporate officers is pretty much a myth in these days of corporate diversification, decision makers do tend to communicate with their counterparts at companies in the same or related industries. They would be far more likely than a foundation or a government funding executive to pick up the telephone to check out your reputation before considering your proposal. So an "in" with just one corporate executive may open the door to others.

Begin in your own backyard. Does someone in your family, a friend, or neighbor have a fairly high position in a corporation in which you have an interest? If not, do you know someone who knows someone at a corporation? All you need is one well-placed individual who can provide you with a letter of introduction or make a phone call in your behalf. Your goal should be to get to see an important decision maker, in person, in order to make the argument for supporting your idea. Unlike your approach to other funders, this meeting generally takes place *before* submission of a formal proposal.

Making the initial contact is the hardest part when approaching corporate funders. Once you actually have secured an interview, you easily should be able to ascertain how good your chances are. Since corporations rarely use professional proposal reviewers, the person with whom you meet might well be the one who will review your proposal and make recommendations on it. He will indicate to you whether or not you need to fill out an application form, whether you should submit a free-form proposal and what it should include.

Your actual proposal to a corporation should follow the basic format outlined in Chapter 7. The introduction immediately should confront the issue of what the corporation stands to gain in supporting you. The needs, objectives, and procedures sections (see Chapter 7) may be shorter than in a foundation or government proposal, but the budget should be fully drawn up, since the corporate officer will probably turn his attention here first. Be sure the budget is designed to emphasize cost-effectiveness and the maximum projected results of your idea, if funded. In the absence of other guidelines, get your proposal in by September, since that's when many corporate budgets are drawn up. Some corporate funders may have money left over at the end of their fiscal year in various departments to disburse for charitable purposes. However, why wait until then?

DECISION MAKERS

If you do secure an interview with a corporate contributions officer, try to bear in mind his own personal motivation and the particular pressures upon him. Why is he working in this area of corporate activities that is usually lumped under "overhead" as opposed to the truly important business of making money? To whom does he report within the corporate structure and what is he supposed to accomplish? It is a safe bet that funding programs and other community relations activities usually are regarded as marginal by the corporation's board of directors. In times of stringent corporate economy measures, these programs will be the first to go. The position of Philanthropy Director has been called: "the kiss of death in most companies."[19(p.41)] Hence, the corporate giving officer may be quite insecure as to his future. His own prestige and self-image may be on the line. These are important factors to consider prior to an interview.

As already noted, since many ideas for support of individuals' projects will fall somewhere else within the corporate structure rather than under the province of the company contributions division, you will have to act as a sleuth to find out who the important decision makers really are. Cultivate any personal contacts you are able to make. This kind of behavior is distasteful to some grant seekers, who view it as apple polishing and beneath their dignity. Nevertheless, it is absolutely essential in order to break through the veil of secrecy at many companies and to find out who really has the power to help you get the support you need.

THE REVIEW PROCESS

One corporate contributions officer recently calculated that requests for grants were arriving at a rate of one every 26 seconds.[22] Although this calculation may reflect a feeling of overload on the part of the officer himself, it is also an indication of some of the pressures being placed on corporations as a result of federal funding cutbacks and current policies. As noted earlier, it is unreasonable to expect corporations to pick up all of the funding programs dropped by the Reagan administration. On the other hand, corporations *are* increasing their contributions and the attention paid to their giving patterns. It is likely that with increased competition for corporate funds, the review process employed by the decision makers who determine who gets what will also receive more scrutiny.

In a recent survey conducted among more than 1,000 top companies listed by *Fortune* magazine, findings about the review process supported the results of earlier surveys: the timing of requests for funding is all important; many of the final decisions regarding corporate funding are made by chief executives rather than by committees; and familiarity with company policies and, in particular, with key personnel (even if it doesn't actually determine whether or not you receive support) does help to speed the review process along.[23]

The review process that proposals to corporations must undergo has been described as a series of "weighing and balancing." Both qualitative factors, like visibility in the community, and quantitative factors, such as profit and loss figures, are at play. There are a number of elements that have an impact upon corporate giving, including company philosophy and policies, size and number of employees, and comparison with companies of similar size in the same or a related industry. When examining your proposal to a corporation, reviewers will question its cost efficiency, its potential impact on the local community, and the extent to which your project benefits employees or others who would not get help otherwise.

Standard reviewing procedures regarding corporate grants are largely irrelevant to the unaffiliated individual grant seeker, since much of the support you eventually receive will come out of public relations, advertising, or other funds. The final decision will most likely be made by a key executive based on one or several of the factors mentioned earlier. (See: Why Do Corporations Give?)

INDIVIDUAL DONORS

FACTS AND FIGURES

We will discuss individual donors briefly, even though outright grants from one individual to another are relatively rare. Moreover, the two main ingredients in receiving support from an individual donor are variables that no grant seeker can control completely: (1) being in the right place at the right time and (2) knowing the right people. On the other hand, there *are* instances of wealthy donors sponsoring unrelated individuals either by providing a place for them to work, buying their work, introducing them to other wealthy patrons, helping them get their work publicized, or by financially supporting them outright, sometimes for years.

Individuals during 1981 donated the major share of U.S. private philanthropy, 83% or approximately $44.5 billion (compared to the $2.6 billion donated by foundations, the $3 billion donated by corporations, and the $100+ billion federal human services budget).[17] As in previous years, more than half of individual giving went to churches and other religious organizations. Individual donors traditionally have not been particularly promising funding sources for the unaffiliated grant seeker, because, although individuals may deduct up to 50% of their adjusted gross income for contributions to charity, gifts to other individuals do not qualify. The charitable deduction allowance is a prime motivator for individual giving. However, the grant seeker should not discount individual donors completely as potential funders. There is always the possibility of affiliating with a qualifying sponsor, especially a religious institution of some sort.

At least where the individual artist is concerned: "The private person is still the largest and most generous patron in America, and his largess is not reported anywhere in a formal way."[24] Here lies another reason to at least consider individual donors as a potential funding source, *if* you can find a way to tap into their motivation. Since gifts from one individual to another are not often publicized (they are frequently kept secret) and not categorized or publicly reported in any formal way, there may be a good deal of this activity going on that we don't really know about. In times of financial and other crises, individuals have been known to rise to the occasion. As with corporations, therefore, we can expect an increase in individual philanthropy during this period of federal cutbacks.

"THE MILLIONAIRE"

A favorite television program when I was growing up was "The Millionaire." Each week an eccentric billionarie gave away $1 million anonymously via his lawyer to a perfect stranger. The recipients were not selected at random but according to private reasons the wealthy eccentric never explained. Usually the recipients were in some sort of financial distress at the time they received the money. The only string attached was the requirement of absolute secrecy. The recipient was not allowed to divulge the source of the money upon penalty of returning the balance if he did. This secrecy requirement invariably led to the recipient's downfall, so that at the end of the program he almost always had no money left—but he had somehow learned a valuable lesson.

I mention "The Millionaire" because it says a great deal about the mythology surrounding individual donors. The reason this television program was so tantalizing was that it corresponded to a daydream that we have all had at one time or another. In this dream, a wealthy individual, much like a fairy godmother or a genie in a bottle, appears before us and gives us a sizable sum of money to do with as we wish. There are no strings attached. The donor approaches *us* so that we don't even have to work hard to seek out the money. The mysterious nature of the anonymous donor absolves us from any feelings of obligation or guilt.

The situation just described is only a dream. There *are* still potential donors who could wipe out a deficit with the stroke of a pen. There are individuals so wealthy that a substantial gift to you would hardly be noticed. They could help you if only they would. However, the fact is that in most cases they won't. Why should they?

WHY INDIVIDUALS GIVE

Before approaching individual donors, perhaps more than with any other category of funder, you must consider their motivation. Since they receive no tax deduction for gifts to other individuals, there must be some alternate reason for them to give. The donor may take an interest in the recipient's work or in him personally. The donor may be looking for a cause to increase his own feelings of self-esteem or to add to his reputation as a supporter of whatever the recipient does. He may simply wish to appear to his friends and family as a generous person. He may be a frustrated artist, performer, scholar, or inventor himself and wish to share vicariously in another individual's professional success.

Individual donors give out of pure generosity or for negative reasons, such as the desire to create a sense of indebtedness in the recipient, the craving for publicity, or feelings of guilt or even fear. The current rise in charitable giving among corporations and individuals to some extent may be "fear inspired."[25] Because of expected federal cutbacks in many education programs and social welfare services, the individual donor may say to himself: if I don't support this or that program, it might well go under.

SOME EXAMPLES OF INDIVIDUAL DONATIONS

Percy Ross, 64-year old millionaire entrepreneur and eccentric, stood on a Fifth Avenue street corner one recent Christmas and handed out silver dollars to passersby. His reason for doing this reportedly was to encourage support of the Salvation Army. Recipients of silver dollars were then supposed to turn around and give paper ones to this charity. No one counted, but the amount given away by Mr. Ross that day was said to be between $5,000 and $10,000, much of it to children. (The Salvation Army did not disclose how much was returned to them in the form of donations.)

James Michener, whose novels have sold in the millions, wanted to help beginning writers and to "channel the surplus back into the profession."[26] With his gift of $11,500, he became the first donor of the National Poetry Series, which sponsors at least five new books of poetry per year. Michener also awarded a huge gift of $500,000 to the University of Iowa Writers Workshop to establish an endowment to support promising young writers during the difficult period before their work is first published. These are examples of gifts from one individual to others, via a sponsor. Mr. Michener was quite generous and got a tax deduction to boot.

I'm sure you've heard of numerous instances where one individual with a way-out scheme or a revolutionary invention finds a wealthier individual to support him, sometimes to the eventual chagrin of the donor. However, who's to say whether or not the donor gets a vicarious thrill, even if the project fails. At least he has the satisfaction of knowing that he took the chance to try to come up with something potentially great. In such an instance the risk taking may well be the motivating force behind the giving away of the money.

Then there are those individual donors whose names you've heard associated with famous prizes that are awarded to individuals: Pulitzer and Nobel come to mind. Other individuals have donated fellowships of lasting benefit to many individual recipients. Rhodes and Danforth are examples of these.

Wealthy individuals give endowments to various universities to benefit particular students and professors. If you go to Yale, and you happen to be named Leavenworth or Deforest, you get a $1,000 scholarship.

In Greenwich, Connecticut, where I live, a cocktail party was held as a fund raiser for a local poet who plans to publish her latest work herself. Patrons paid $15 each to attend the party and to hear the poet read selections of her poetry. In this way, members of the public were supporting an individual artist.

All over the country, similar events, dinners, and exhibitions are held for photographers, sculptors, painters, and others. Sometimes the price of admission is quite steep. Often the amount is left unstated, but those attending are expected to bring along their checkbooks. Wealthy individuals attend, become interested in the artist, introduce him to their friends, and help him to get his work commissioned, purchased, and displayed. Similar benefits are held for individuals who are promoting various causes. Some are affiliated with nonprofit sponsors; some are not.

The above examples are indicative of the varied ways in which individual donors support other individuals. They also exemplify the diverse factors that motivate an individual to give money to support another individual's work.

WHAT THE DONOR WILL EXPECT IN RETURN

Whatever his reasons for supporting another unrelated individual, one can be sure that a gift from an individual donor, even if "anonymous," usually comes with strings attached. In fact, individual donors are the most likely of all funders to expect something specific in return. At the very least, they will probably require a show of allegiance and/or gratitude. They may even want to change your ideas in some meaningful way to correspond to theories or plans of their own.

As far back as the Medicis, the famous Renaissance patrons of the arts, individuals received support from wealthy patrons. However, something was always expected in return. One expert on arts funding warns that "loyalties, works, amusements or other satisfactions from the artist" will be expected as a matter of course from private patrons in return for their help.[24]

If you are an artist or otherwise, think carefully before you accept support from an individual donor. Compromising your integrity as a professional in whatever field you are in is a possible occurrence unless you are careful. Always consider what price you may have to pay. This may not be stated explicitly (and you don't want to appear ungrateful), so you will have to do some subtle detective work to find out. Trust your own instincts. If a gift proffered doesn't feel right, don't accept it. Remember the case of "The Millionaire" in which an anonymous gift with apparently few strings attached always proved impossible to keep.

APPROACHING INDIVIDUAL DONORS

Throughout this discussion I have used the term *gift* rather than grant because the term *grant* suggests a formal approach or application process and some form of supervision. None of these may pertain in the case of a gift from an individual donor. Whom you know and your own personal reputation will be most helpful here. In some very rare cases, just like the genie in the bottle, the donor will approach *you* (usually via a mutual acquaintance) to see if you need help. Ordinarily, there will be no written proposal and nothing in writing at all. Simply a check or checks will be made out to you. The donor will base his decision on his own appraisal of your work or ideas and of you as a responsible individual.

Although it is absolutely true that knowing the right people will greatly enhance your chances of receiving support from individual donors, this does not mean you must sit back and be passive. There are ways of researching wealthy individuals (see Chapter 6) and of placing yourself in their paths. There is no directory of people who have money to give away, so luck plays a major role in your matching yourself up with an individual donor.

Often the donors themselves don't even know they want to make a gift until they meet you, examine your work, or are otherwise exposed to your ideas. As with other forms of funding, a gift from one individual frequently leads to further support from others. To increase your chances of receiving such support, see to it that your accomplishments are well publicized, talk to people, keep your eyes and ears open, and, above all, explore your present contacts.

Finally, as with corporations, consider the forms of nonfinancial support you might receive from individual donors. These other kinds of support may prove to be worth more than an actual cash gift. Do not underestimate what a wealthy, influential supporter can do for you. If you're an artist or writer, he may have important contacts, access to the media, prestigious reviewers, or purchasers of work such as yours. If you're a scientist or inventor, he may be able to promote your idea, to introduce you to others who can, and even to help you secure grants from other types of funders. Individual supporters can also be very helpful in finding you a place to work. Simply talking up your idea among his friends may result in your receiving all kinds of support, cash and otherwise. One important contact with a wealthy, influential, or prestigious individual donor, patron, supporter, or mentor can lead to a lifetime of successful grant seeking.

An Example of a Bequest

Our discussion thus far has centered on living individual donors. However, bequests also may be sources of funding for the individual grant seeker. One further anecdote will serve as an example of the often unusual ways funds from wealthy individual donors may be made available to those who seek them out, even though the existence of such funds is not advertised to the community at large, and luck still plays a major role in hearing about them in the first place.

In my hometown of Marblehead, Massachusetts, a man died and left several hundred thousand dollars to build a church. For one reason or another the church was never built, and the money, reasonably invested, accrued to about $1 million. The trustees of the estate went to court to alter the will. Since the deceased individual was known to have had an interest in the arts and in medicine, the bequest was changed to benefit young people planning to enter these fields. The interest accrued on the money then became available as scholarships. Any recent graduate of that town's high school entering art, medical, or dental school was eligible. Interestingly enough, eligibility was based on these requirements plus scholastic ability and not on financial need. Quite by accident, a member of my family heard about the scholarship, although it was not then well known in town and the school guidance counselors were not dispensing information about it. This relative applied to the trustees of the endowment and was accepted. Subsequently, he attended medical school for four years with his full tuition and many of his expenses paid from the interest on the bequest.

A rare example? Maybe not. It is possible that such funds are available in many hometowns. You might be able to benefit from this type of support if only you have the good fortune to be in the right place at the right time to hear about it and then the fortitude to go out after it.

MISCELLANEOUS FUNDERS

This truly is a catchall category covering a wide range of institutions and groups as diverse as the American Legion and local historical societies. The following organizations, in the proper circumstances, might make awards to individuals: clubs, arts and other councils and societies, trade and professional associations, federations, research institutes, Greek letter societies and fraternal groups, religious bodies, chambers of commerce, veterans organizations, museums, hospitals, and even libraries.

In addition, bar associations, labor unions, and service clubs like local Junior Leagues are potential sources of funding for various individual projects. Groups like the Rotary, Lions, or Elks clubs often give annual awards or scholarships to young people. Churches and other denominational groups represent a relatively unexplored source of funding for individuals and grass roots groups. Various special interest groups sponsor annual competitions, for which the prize money may be substantial.

DEFINITION OF TERMS

In the discussion of individual and some corporate donors, I used the terms *gift* or *donation* rather than grant, because many of the processes of identification, application, review, evaluation, and supervision normally associated with seeking a grant do not pertain. For most funders in the miscellaneous category I'll refer to *awards* or *prizes* rather than grants.

The editors of *Webster's New World Dictionary* (2nd college ed.) offer the following definitions (the italics are mine):

Grant or grant-in-aid: "a grant of funds, as by the Federal government to a State or by a foundation to a writer, scientists, artist, etc., *to support a specific program or project.*"

Award: "to give as the result of *judging the relative merits* of those in competition (to award a prize for the best essay)."

Prize: "something offered or given to *the winner of a contest.*"

As you can see from these definitions, the distinctions among the terms *grants, awards,* and *prizes* are a question of emphasis. A grant consists of a gift of cash to bring to fruition a particular idea. If you receive a grant, the assumption is that you went through a certain process, that is to say, you worked for it. Grants relate to specific activities you will perform at a future date and are often (in theory at least) renewable. Awards may be cash, but they often include or may consist entirely of placques, medals, certificates, and the like. If you receive an award, it is a form of recognition either of one specific accomplishment on your part or of the promise your body of work represents in a given field. Awards generally are one-shot deals; individuals may apply for them, or they may be granted only through nomination. Some funders in the miscellaneous category offer prizes. If you receive a prize, the assumption is that you won it in a contest, that is to say you beat out the competition. Luck may also play a role in your winning a prize.

Funders in the miscellaneous category generally give out awards or prizes rather than grants (according to the above definitions); these tend to consist of small sums of money. However, some of the funders also give scholarships, fellowships, and research grants.

How the Individual Can Benefit from Awards Given by These Funders

If you are of a particular racial or ethnic background *or* if you, your brother, father, grandfather, or uncle belongs to a particular fraternal organization or society, there is probably an association that may benefit you through scholarships, awards, or help with career development.

If you are associated with a particular cause or your grant idea serves to promote that cause, there may well be an advocacy group or lobbying organization associated with your field of interest that will either award you a small grant or put you in touch with those who can.

Do you excel at a particular sport? You may be able to enter a competition or find a sports group to sponsor you, award you a scholarship, or help advance your professional career.

Do you or a member of your family belong to a labor union, league, or guild? Most unions maintain funds specifically for the benefit of individual members and their families to use especially in times of medical or other financial crises. You can also borrow against such funds at low or no interest for educational or professional development. Some unions offer outright grants or sponsor research in areas related to workers' interests. Union officials can help you make contact with others who might offer financial support. Invariably, these services are "for members only."

Veterans, hereditary, and patriotic organizations award scholarships, loans, small grants, and prizes usually to recipients in particular geographic locales. You may qualify by dint of your ancestors or because of some proficiency on your part.

Do you belong to a Greek letter society? Those that are solvent offer awards, scholarships, and fellowships to members in good standing.

Trade, business, and commercial associations give achievement awards, most often in the form of medals, plaques, or other noncash recognition of accomplishments. However, they also give small research grants for matters relating to the promotion of their business interests or to help develop products or services.

Scientific and technical organizations and research institutes offer residencies, internships, and fellowships and serve as sponsors of individual research projects. These awards tend to be highly specialized with many eligibility requirements.

Cultural organizations and local agricultural and historical societies conduct competitions, sponsor exhibitions, and offer prizes. Some make awards and offer fellowships. Local talent is preferred. Cash amounts tend to be quite small.

Health and medical organizations, like other scientific and technical institutions, offer research grants and fellowships to qualified professionals, especially to those who already have a "name" in their field. These grants may be substantial.

Almost any disease you can name has its own association. If you specialize in research on the causes of or cures for a particular disease, or if you are unfortunate enough to suffer from the disease, you may be eligible for a research fellowship or prize.

Is your grant idea related to a hobby? Whether you perform a craft, collect Wedgwood, or play chess, collectors' groups exist that make awards, sponsor research, and offer educational programs, festivals, and exhibits in your field of endeavor. Many have annual achievement awards for which you might qualify.

How do you find out about all these various groups? The two main sources of information, "The Invisible Network" and *The Encyclopedia of Associations,* are described in Chapter 4. Once you identify a particular funder in the miscellaneous category, the rest is up to you. You will have to call or write and probably do a good deal of leg work, such as gettting appropriate endorsements and nominations, filling out detailed forms, appearing for interviews, and so on. However, take heart, because of the public relations function that many of these associations and special interest groups perform, funders in the miscellaneous category have the highest response rate. They are likely at least to answer your letter or phone call, to tell you whether or not you qualify, and to send you their guidelines and application forms.

SOME MORE EXAMPLES

Here are some specific examples of awards and prizes given to individuals by various associations and groups in the miscellaneous category over the past few years.

The $100,000 Lila Annenberg Hazen Award for excellence in clinical research in medicine recently went to a Rochester University immunologist. The winner of the award must choose a research fellow, and the immunologist chose a resident at Kings County Hospital. The award is administered by Mt. Sinai Medical Center.

The 1981 Booker Prize, a British award of $5,500 for best novel of the year, went to the author of the book *Midnight's Children* (New York: Knopf).

The first recipient of the William Schuman Award, a $50,000 prize to an American composer for lifetime achievement, was announced by Columbia University, its administrator. Henceforth, the award will be given out approximately every 2 years.

In 1977, Jan Mitchell, art collector and former owner of Lüchow's Restaurant in New York, established the Mitchell Prize, which recognizes the year's most distinguished contribution in English to art history. In 1981, the prize of $10,000 went to a British writer. In celebration of the prize's fifth anniversary, an additional $2,000 was awarded to the author of the most promising first book of art history: *Leonardo da Vinci: The Marvelous Works of Nature and Man* (Cambridge, Mass.: Harvard University Press).

The 33rd annual New England Exhibition of Painting, Drawing and Sculpture was held at The Silvermine Guild of Artists in New Canaan, Connecticut, in May 1982. Thirty-six works were designated for awards in the show. Prizes ranging from $150 to $1,000 are annual awards donated by local corporations to New England artists in various media and administered by the Guild.

Rotary International offers graduate fellowships, scholarships, and other awards, the purpose of which is to promote international understanding. Recipients may be students, technicians, journalists, or teachers of the handicapped who become "unofficial ambassadors of good will." Among the many eligibility requirements, you must find a local Rotary Club to serve as your sponsor.

The Carnegie Hero Fund Commission in Pittsburgh, Pennsylvania, gives medals and awards to individuals in recognition of selfless acts of heroism, to persons injured in such acts, and to widows and orphans of deceased heroes. Awards are made through a nomination process and after completion of a formal application procedure and interview. In 1980, $306,000 was awarded to 242 individuals, the average award being around $1,500.

The Society of Medical Friends of Wine in Sausalito, California, offers a $1,000 award every other year for original published research on the nutritive and therapeutic values of wine.

South Paws International of Birmingham, Alabama, gives a "Southpaw of the Year" citation to an outstanding public figure who is left-handed.

These awards and prizes are as diverse as can be. Although the various groups donating and administering these types of funding appear to have very little in common with one another, several general rules must pertain. I set out to find out what the areas of commonality might be by conducting my own informal survey of funders in the miscellaneous category.

My Own Survey

During 1981, I contacted approximately 100 miscellaneous funders by letter, requesting information on their awards, prizes, or grants to individuals for this book and promising them anonymity. This last promise was made in order to encourage them to respond. As already noted, funders hesitate to provide information if the result might be a flood of new applicants. The 100 organizations were selected from the *Encyclopedia of Associations* according to several criteria: because their names were familiar as prestigious funders of individuals, because the description of their activities appeared of special interest to individuals, or simply because the name of the organization was intriguing. An attempt was made to cover various categories of associations, giving awards in a wide range of subject fields and to different types of individual recipients.

About one-half of those contacted responded with information on awards for individuals. Anyone with experience in direct mail requests for information from funders will recognize that this is an extremely high response rate. A majority of the letters were signed by "Executive Directors," a large number of whom were women. The following information was gained from my survey.

Those funders in the miscellaneous category with established awards programs often use outside specialists to review applications, but final decisions are almost always made by a committee or by a board of trustees or directors. Two styles of approach are favored: For achievement awards or prizes, nomination from a prestigious individual or institution is enough, often with no actual proposal letter involved; for scholarships, research grants, and competitions, formal application forms with detailed personal data must be submitted. Deadlines are strictly enforced. Both approaches usually are followed by personal interviews, if your application is to receive serious consideration.

You should expect an award from a miscellaneous category funder to be quite small. Awards of $500 to $1,000 are typical. National associations may require sponsorship by a local chapter before they will fund you. Geography often plays a major role in the giving out of these awards; local talent almost always is preferred. As is usual with funders, knowing someone on the board helps. Some awards are only for members of the various associations and clubs. Others have restrictions on how you can use the money. For example, an organization called Religion in American Life gives an annual award of $5,000, but the money must then be donated to the recipient's favorite charities.

When I discussed foundations, I advised you to keep in mind that you were interested only in those that had endowments of their own (or received annual gifts from the donors) and that actually make grants to individuals. Many of the funders in the miscellaneous category are quite actively soliciting funds from the general public, their membership, special interest groups, and wealthy individuals. Traditionally, those groups seeking funds themselves are not considered as likely targets for the grant seeker, because when times are bad and competition for these awards increases their funding disappears. Their boards may decide not to make awards for several years at a stretch. If you are applying for an award, check on the financial solvency and fund-raising activities of the organization involved.

TYPES OF AWARDS

Achievement Awards and Prizes

Such awards usually are made via a competitive nomination process. Generally they are in recognition of a stock of knowledge, body of work, or a specific accomplishment on the recipient's part. Writers, scientists, artists, performers, and civic-minded individuals would be most likely to benefit. These achievement awards sometimes called "merit awards" are usually quite small. In some cases, there is no cash at all, simply a medal or certificate conferred upon the individual. Most achievement awards require formal nomination, although in some instances nomination can come from almost any source. If you think you qualify for an achievement award, it is possible to seek out an institution or individual to nominate you. You can even handle all the paperwork yourself.

Competitions

Funders in the miscellaneous category also sponsor competitions. These are generally of two types: contests open to all candidates with very specific rules under which you must apply and those for which you must be nominated, recommended, or invited in order to participate. In the former case, you can expect competition among applicants to be quite stiff even though there may be little more at stake than formal recognition of one's work. In the latter case, familiarity with the politics of the nomination process is most helpful. In some instances, unfortunately, the reputation and professional clout of your nominator will have more to do with the outcome than your own qualifications.

Some competitions open to all, such as drawing, essay, poetry, limerick, and logo contests, involve very small cash awards. Such competitions, furthermore, may entail entry fees, which are really fund-raising gimmicks. Check out these competitions carefully before you waste your time applying.

For the second category, those competitions for which you must be nominated or invited to participate, the cash prizes may be substantial. Examples of these are competitions for architects to design or redesign particular structures and scholarships for students, scientists, inventors, or computer programmers. There are lots of rules for these competitions, which must be followed minutely or you will be disqualified.

Awards as Small Grants

I've already mentioned that some funders in the miscellaneous category use the term "award" as synonymous with grant. In fact, some of these organizations do give small amounts of money to artists, inventors, researchers, or other individuals to perform specific tasks or to complete particular projects. Such grants quite often are at the discretion of the executive director of the association. There is usually little formal application process other than a letter and a personal interview. Once the grant is made, scant supervision may be exercised by the funder. However, with this kind of grant quite often there are strings attached.

As always, you must consider the motivation of the funder. It is not unusual for the publicity or public relations value of the grant to be a prime motivator in the awarding of the money. As the recipient, you may be required to perform certain services, appear in public at various functions, or otherwise publicize the largess of the funder.

Such cash grants often have highly specific eligibility requirements. For example, there is an association that makes grants to needy cartoonists. If you fit all the requirements, then you should have smooth sailing in actually getting the money. Others have many restrictions on how you can use the money. Investigate these limitations carefully before you accept such an award.

Many funders in the miscellaneous category give out scholarship awards. This may be their sole function, or they may do it in addition to achievement and other awards programs. A typical pattern is the donation of from one to twenty scholarships or fellowships per year in small amounts of money to members, those affiliated with the association in some way, or those whose work may ultimately benefit the field. Such scholarships may not be well known to school guidance counselors, and word of mouth is most helpful in hearing about them. Also check bulletin boards at various educational institutions and libraries for announcements of their availability.

Noncash Awards

As already noted, many of the awards from funders in the miscellaneous category involve little or no cash at all. Nonfinancial awards given out by the 100 associations in my survey group included medals, plaques, certificates, citations, trophies, medallions, charms, and other jewelry. There is even an association that gives out a ceramic bust of Edgar Allan Poe as its award for the best mystery novel, short story, or film of the year.

Other organizations do not actually make grants but perform in the role of sponsors for individual grant projects that require institutional affiliation (see Chapter 2). Still others have developed creative ways for serving as clearinghouses or middlemen to help talented individuals further their careers. Some, like Affiliated Artists, have residency programs or sponsor internships for individuals. One organization called The Center for Field Research serves as a screening and coordinating agency that arranges both the funding and the volunteer support for archeological expeditions. Volunteers are individuals with an academic or avocational interest in a particular discipline and a desire to learn by hands-on involvement. However, there's a unique twist to this setup: *You* provide the funding (if enough "volunteers" put up cash, then there will be an expedition); they provide the professional staff and the opportunity to go on a real dig.

Some of the organizations in the miscellaneous category may not give grants, but they will pay your travel expenses to attend conferences, conduct research, present papers, participate in seminars and symposia, and so on. Expenses to attend awards ceremonies are quite often covered by the benefactor. These are valuable fringe benefits, since many of the traditional funders exclude travel funds from their programs of grants to individuals.

WHY BOTHER WITH FUNDERS IN THE MISCELLANEOUS CATEGORY?

Such funders are hard to identify, there is much red tape involved in applying to them, and potential cash rewards seem, for the most part, to be quite small. Also, it must be pointed out that since a number of these awards recognize achievement on the recipient's part, they are more likely to go to established professionals than to those individuals just starting out in their chosen fields. These kinds of awards, nevertheless, are nice—if you can get them.

The prestige factor involved in becoming the recipient of one of these awards may be far more important to the individual than the amount of cash (if any) received. Recognition by one's peers and other experts in one's own field provides an inestimable boost to one's own work, both in terms of morale and potential future financial support. For example, the "Connie" Award may not be familiar to outsiders, but the individual who receives it each year is acknowledged with respect and envy by those in the fields of conservation and ecology. Applying for and receiving such prizes and achievement awards serve as the test and proof of your professional work and enhance your position within your own field. To quote one award-seeking photographer: "Becoming a grant recipient is a permanent distinction like Captain, General, Senator or other similar designation."[27]

Awards from funders in the miscellaneous category entail a certain amount of exposure, which at least for some individuals makes it all worthwhile. When you receive an award, there is usually a degree of attendant publicity for you and your work in the print and broadcast media. The end result could be sales, jobs, commissions, valuable contacts, research contracts, and even larger grants. If you are an artist, craftsperson, or inventor, participating in various exhibits or festivals with prizes provides the valuable opportunity of having your work viewed by major patrons and establishing contact with potential wealthy individual donors.

It almost goes without saying, finally, that awards look good on your resume and on your grant application to major funders. I mentioned at the very beginning of this book that you may wish to use awards, honors, and small grants you receive as stepping stones to trade up to a larger grant. In an article on "The Grants Game," one writer points out that "an impressive string of smaller grants can be a big help when you're ready to move up to the big league."[28]

At that time, you will probably need to affiliate with a sponsoring organization. For now, keep in mind that no matter how unsubstantial in terms of actual cash, awards of this type are regarded by proposal reviewers and other funding executives as concrete evidence of professional recognition of your capabilities and society's support of your work.

NOTES

[1] Nielsen, W. A. Needy arts: Where have all the patrons gone. *New York Times,* October 26, 1980, Section 2, pp. 1; 26–27.

[2] Amis, K. Government shouldn't fund the arts. *New York Times,* August 31, 1980, p. E15.

[3] Schonberg, H. C. New York State's share of arts and humanities grants increases. *New York Times,* May 17, 1981.

[4] Straight, M. *Twigs for an eagle's nest.* New York: Devon Press, 1979. P. 170.

[5] Stop funding artistic circuses. *New York Times,* April 15, 1982, p. A22.

[6] Lee, L. *The grants game.* San Francisco: Harbor Publishing, 1981.

[7] Orr, S. Coping with bloc grants. *Grantsmanship Center News,* November/ December 1981, p. 40.

[8] Glueck, G. New endowment head explains grant reductions. *New York Times,* February 21, 1982, Arts Section, pp. 1; 26.

[9] Kurzig, C. M. *Foundation fundamentals.* New York: The Foundation Center, 1980.

[10] *The foundation directory* (8th ed.). New York: The Foundation Center, 1981. P. x.

[11] *The foundation directory* (6th ed.). New York: The Foundation Center, 1977. P. xi.

[12] *Foundation grants to individuals.* New York: The Foundation Center, 1977. P. i.

[13] *Giving USA; 1982 annual report; a compilation of facts and trends on American philanthropy for the year 1981.* New York: American Association of Fund Raising Counsel, 1982. P. 80.

[14] J. Coleman as quoted in Teltsch, K. Costs and rules putting pressure on foundations. *New York Times,* April 15, 1980, pp. B1, B12.

[15] *Grantsmanship Center News,* November/December 1981, p. 7.

[16] Troy, K. (comp.). *Annual survey of corporate contributions.* New York: The Conference Board, 1981.

[17] Boyd, T. The business of business is . . . philanthropy? *Foundation News,* May/June 1982, p. 57.

[18] Hillman, H. *The art of winning corporate grants.* New York: Vanguard Press, 1980. Pp. 5–9.

[19] Shakeley, J. Exploring the elusive world of corporate giving. *Grantsmanship Center News,* July/September 1977, pp. 35–59.

[20] Carr, E. G. & Morgan, J. F., *Better management of business giving.* New York: Hobbs, Dormon, 1966. P. v.

[21] Bishop, J. E. Foundation raises money for cancer research. *Wall Street Journal,* March 20, 1981, p. 48.

[22] W. A. Orme as quoted in Teltsch, K. The challenge on donations. *New York Times,* November 4, 1981, p. D11.

[23] Murphy, D. J. *Asking corporations for money.* Port Chester, N.Y.: Gothic Press, 1982. Pp. 30–31.

[24] Millsaps, D., and the Editors of Washington International Arts Letter.

National directory of grants and aid to individuals in the arts (4th ed.). Washington, D.C., 1981. P. iv.

[25] Teltsch, K. Gifts to charity rise 12.3% to 53.6 million, a record. *New York Times,* April 19, 1982, p. 16.

[26] Brewer, D. Throwing dice for writers. *Coda,* April/May 1981, p. 9.

[27] Deschin, J. Want a photographic grant? *35mm Photography,* Winter 1973, p. 16.

[28] Bellows, B. The grants game. *Writers' Digest,* January 1981, p. 37.

"Your dream, of course, may be to ... go find funding (somewhere) that will enable you with singularity to do the one thing you most want to do."

—RICHARD NELSON BOLLES
What Color Is Your Parachute?

"Modern fundraising is 90% research, 10% solicitation."

—TAFT CORPORATION
promotional literature

Chapter 6 / FINDING THE RIGHT FUNDER FOR YOU

DOING YOUR HOMEWORK

According to one experienced financial aid officer, "millions of dollars worth of scholarships, grants, and other financial aid go unclaimed every year."[1] In 1979 alone, $135 million in possible grant funding was not awarded simply because people didn't know where to look. Identifying potential funders, or what seasoned fund raisers refer to as "prospect research," is a lengthy, tedious process. Although it lacks glamour, there is no way of getting around the essential detective work involved in finding the right funder for you.

Prospect Research—a Time-Consuming Process

As with many other elements of grant seeking, finding the right funder takes time. However, it is time well spent, since there is no substitute for good information that is well maintained and easily accessible.[2] This kind of research is really a narrowing-down process. You start with the broadest possible base and slowly winnow out those funders whose priorities or limitations exclude you or your idea or who appear inappropriate for other reasons.

You may uncover prospects in a variety of ways, including extensive research in reference books, long hours spent at the library, word of mouth at a meeting or conference, recommendations from mentors, colleagues, or other grant seekers, or even mere happenstance. However, this initial uncovering of the name of a potential funder is just the beginning. It is absolutely essential that you learn as much as possible about your potential funder *before* you submit a proposal. This is what "finding the right funder for you" is about. There is, in fact, a direct correlation between the amount and accuracy of information you receive on a given funder and your chances of success. The less information available on a funder, the smaller the likelihood you will get a grant.

In addition to a lot of accurate information, you need the most current data available. Always ask yourself if more up-to-date information might be secured. Although the basic giving patterns of most funders are more or less stable, private foundations and corporations sometimes have subtle shifts in priorities and changes in management that affect their giving policies. State, local, and federal government funders are subject to extreme shifts of emphasis when new administrations take over, agencies change hands, and bureaucrats are hired or fired. It is essential that you find out about these changes so that you don't waste your time researching an obsolete grants program.

When approaching a published source of any kind, first look at the date. Be very wary of anything more than two or three years old (especially regarding government funding) as to specifics on grant money available. On the other hand, for general recommendations on trends and styles of approach as well as for materials on proposal writing and grant administration, the fact that a publication is old does not necessarily make it out-of-date. In my own research for this chapter, I often found the prefaces, introductory materials, and appendices of publications that sometimes were ten or more years old to have thoughtful and insightful discussions on grant-seeking methods and application procedures, which really will never become obsolete.

RESEARCH METHODOLOGY

In your quest to identify potential funders you might adopt the method used by professional fund raisers to solicit individual donors called *external cultivation,* or finding out all possible information that is relevant. Under this procedure, the grant seeker performs in three roles: (1) *source collector*—collects valuable information from libraries, reference books, computer files, tax returns, news clippings, word of mouth, bulletin boards, and so on; (2) *source rejector*—eliminates inappropriate funding agencies; and (3) *source selector*—singles out the most likely candidates to fund your idea and rates them in order of priority.

Source Collection

According to Carol Kurzig, Director of Public Services for the Foundation Center, "it takes serious time-consuming research to track down the foundations whose giving records or stated objectives are most directly related to your goals."[3(p.33)] This statement pertains to most other funders, as well as to foundations. During the data collection stage of your research, it is important to note the distinction between what a funder *says* it does (its "stated objectives") and what grants it actually has awarded in the *recent* past (its "giving record"). Many funders purposely leave their stated objectives as broad as possible. You should always refer to current grants lists, described later in this chapter, for more specific information.

Source collection is a twofold task. First, you find out what information sources exist for your particular grant idea; then you find out where and how best to use existing resources. This stage of your research is what Barzun and Graff in *The Modern Researcher* call "the digging and delving" stage.[4(p.x)]

Finding the right funder requires facts. To ascertain that a potential funder actually has money to give away, don't just rely on hearsay. This brings us to the single most important aspect of source collection: verification. Don't accept any information you come up with at face value. Simply because something appears in print does not make it so. By the same token, for information you get by word of mouth, learn to discriminate between report and rumor, what is probable and what is doubtful. This aspect of your research requires sound judgment. If at all possible, try to find a seconding source for each bit of information you read or hear about. In the words of Barzun and Graff, "there is no substitute for a well-placed skepticism."[4(p.110)] All facts are subject to interpretation. So when you approach any information source, ask yourself who's doing the interpreting.

Source Rejection

This is probably the most crucial stage of your prospect research, since it may well determine whether or not your proposal reaches the "right" potential funder. Some grant seekers are just naturally better at this process than others because it relies on one's own intuition. According to Barzun and Graff, "imagination seconded by patience is indispensable to success."[4(p.68)] For the grant seeker researching potential funders, the more you know, the easier it is to imagine where to learn more. However, at some point, you will stop collecting data and begin to review the potential funders about whom you've collected information. At this point you are trying to determine how well your own needs match the funders' interests, or what fund raisers call *finding a match*.

Approach this stage of your research creatively. As you reexamine the data you've collected on funders, look for similarities and differences, patterns and links, especially for grants awarded in the recent past that seem to relate to your own idea. Since it is unlikely that you'll find a record of a grant awarded that is an exact replica of your own project (and if you did, it might well negate the purpose of your grant search), try to find funders whose recent grants are similar in some way to the one you want.

As mentioned earlier, the processes of identifying potential funders and of eliminating inappropriate ones require strong intuitive ability. You must be able to read between the lines—to feel that a prospect is or is not right. This is a talent that the novice grant seeker needs to learn to develop. So don't despair. Although the techniques required to find the right funder take talent and intelligence, they get easier with practice. As you continue to conduct grant searches, you will become more familiar with the resources in your field, and your intuition will simultaneously blossom.

Source Selection

During the final stage of researching your prospects, you decide which funders you will actually ask for a grant. This is also the point at which you will demonstrate to funders that you *have* done your homework. After sustained inquiry and considerable research, you will have selected particular funders who are the most likely to fund your idea.

Don't let the "law of averages" do your work for you. Sending out well-researched proposals to carefully selected funders is a far better procedure than a blanket mailing to funders chosen more or less at random. For most grant ideas, mass mailings simply do not work; they are a waste of time, postage, and money and may give you a bad reputation with funders to boot.

Review and evaluate all the data you've collected with a critical eye. Rate your potential funders in order of priority. Once you've selected funders in which you have an interest, and you've done all your homework on these funders, don't be afraid to *ask* (by phone, letter, or in person) any questions you still have. If you don't understand something or you're missing crucial data, it can't hurt to ask for it. Meekness is not a winning trait in grant applicants. In the absence of complete, accurate, and up-to-date information, you cannot make an informed decision about whom to approach first for a grant.

THE ABCs OF TECHNIQUE

There are a number of helpful hints regarding the mechanics of prospect research that will make finding potential funders somewhat easier. The following are some common research problems along with recommended solutions.

What's in a name? Sometimes you will hear of a particular donor whose foundation or company makes grants for such-and-such a purpose. However, when you go to look it up, you find that no such funder exists.

Keep in mind that for a variety of reasons, many foundations and corporations have names that are different from the original source of their endowment money. To determine if an individual has a foundation or is affiliated with one, see The "Index of Donors" in the *Foundation Directory, Trustees of Wealth,* or *Who's Who in America.* To determine if an individual has a corporation or serves on corporate boards, see *Standard & Poor's Register, Who's Who in Finance and Industry,* and other reference tools discussed in the Corporations section of this chapter.

When researching foundations, you should perform a subject search for larger foundations (if your idea is national in scope) and a geographic search for smaller foundations (if your idea is local in its projected impact), or both simultaneously. You will also require two types of information: where the foundation says its interests lie and what ideas in actuality it has recently funded.

For moment-to-moment shifts of emphasis among funders—be alert to media references (watch the news) and periodical articles, especially columns in magazines, like *Time, Newsweek, U.S. News and World Report, Business Week,* and *Forbes,* and newspapers, like the *Wall Street Journal,* which often provide the most up-to-date insights into current philosophies of giving and funding priorities.

What method of note taking should you use? Should you use yellow legal pads, loose-leafs, spiral notebooks, 3 x 5 cards, tape recorders? The method is up to you. But once you adopt a method, if it works, stick with it. Inconsistent notes in various formats can prove quite frustrating when it comes to selecting your best prospects. Write only on one side of the paper or card. Also, take the time to jot down the complete source of every piece of information, including the date. You can never tell when you'll need to refer back to the original source at a later time.

Most experienced grant seekers find it helpful to make up *data sheets* or prospect worksheets on each potential funder. These sheets come in various formats,* but generally they provide for the most needed information on each potential funder (see next section). Data sheets

---

*For a sample data sheet, see C. M. Kurzig, *Foundation Fundamentals* (New York: The Foundation Center, 1980), pp. 99–100.

allow for a consistent format of note taking, and, like checklists, they help you to determine quickly if there is information still lacking on a given prospect.

View each printed information source you refer to with a critical eye. To test for reliability, first look at the publisher's name. Is it one you've heard of, or is the book issued by an unfamiliar organization or even privately printed? Many fund-raising guides and some directories of funders are issued by organizations not ordinarily in the business of publishing books. Some are really promotional pieces for an agency's own services or other publications; others started out as in-house handbooks that someone decided were good enough to sell. All kinds of groups get into the act, and they provide all kinds of advice to grant seekers. Just because something appears in print does not make it true or worthwhile. So approach the various publications with a healthy dose of skepticism, and take careful note from whence the advice comes before you follow it.

To check for *currency of information* look at the date of publication. Is this a revised edition? Is there a more current edition available? Next, determine who the author, compiler, or editor is. Check for a brief biography on the back cover or flyleaf or a list of other publications credited to the author. Is his name or affiliation familiar to you? Before you begin to take notes, look for a brief section on "How to Use This Book," which might give clues as to limitations, exclusions, resources the authors used in compiling the data, and material that is included and left out. This type of careful initial scrutiny of each publication will save you time in the long run.

Plan your research time. Save the longest stretches of time for the study of your main sources, which should include the reference tools discussed in the next sections. Use shorter periods of time for "broken field running"[4(p.325)]—verifying names, dates, addresses, and phone numbers and hunting down references. Allow enough time; finding the right funder most probably will take longer than you might think.

LIBRARIES—AN ESSENTIAL RESOURCE

In earlier chapters we talked about various types of resources for the grant seeker. Organizations were discussed in Chapter 2, and people, in

Chapter 4. Printed resources will be discussed in detail later on in this chapter. However, an important resource frequently overlooked by the grant seeker is the library.

Throughout this book we have noted that you should always begin your grant search in your own backyard. Your local public library is "the researcher's first port of call."[4(p.52)] So, learn about your library; it helps you to overcome feelings of ineptitude. Attend orientation sessions or tours, and familiarize yourself with the card catalog (a surprising number of people don't know how to use this invaluable reference tool), shelf arrangements, location and access to copying machines, and microfilm reading and reproduction equipment. Ask the librarian for help. Give this professional some background about your grant idea and research needs. You'll be surprised how helpful the librarian might be during this stage in your search for prospects.

Your local public library most likely has many of the standard biographical reference works and information on local funders and on community resources. However, you need not feel restricted by the limitations of your own library's collection. If it doesn't have what you want, you can get many publications through interlibrary loan from other institutions. Most libraries belong to networks that facilitate interlibrary lending, and they have a librarian who will take care of the paperwork for you. If the books you need can't be borrowed from another library (many reference books cannot), relevant portions may be photocopied or microfilmed (subject to the rules of copyright). The same goes for periodical and newspaper articles.

In addition to public libraries, there are various technical and special libraries dedicated to specific subject areas and fields of interest, corporate libraries, academic libraries, state libraries, federal information centers, historical societies, art associations, and museum libraries, as well as private collections. Your local public library will have directories of these. Check *The American Library Directory* and the *Directory of Special Libraries and Information Centers* plus various state and subject directories as well. Refer to these directories to determine which types of libraries might have information relevant to your grant search and whether or not these libraries are open to the public.

The Foundation Center maintains excellent libraries in New York, Washington D.C., Cleveland, and San Francisco with virtually everything in print about foundations and their grants, plus many useful directories and funding guides for grant seekers in numerous fields. Many of the publications discussed in the bibliography section of this chapter are available for your examination at The Foundation Center libraries. The Center also has microfilm copies of the records that foundations file with the IRS each year (usually a year or two behind the present date). In addition to its own four libraries, the Center maintains cooperating collections in all fifty states, in foundation, academic, and public libraries, which have information on local foundations. These are open to the public, and some even provide telephone reference services. Contact The Foundation Center in New York for an up-to-date list of cooperating collections.

For information on corporations, you may wish to visit company libraries (you'll need an appointment), graduate business libraries at universities, or the business division of your local public library. All of these libraries maintain directories, biographical data on corporate executives, corporate annual reports, and other useful information. Some have unpublished reports and studies that relate to local corporate funding and current priorities.

Many states have official state libraries, which provide information on state funding agencies, the state's share of federal funding, and biographies of state officials. Your state budget, available for examination at the state library, provides valuable insight into local needs and priorities and is useful backup material for the "need" section of many grant proposals.

The federal government has designated depositories in approximately 1,200 public and academic libraries throughout the country that receive, maintain, and make available federal studies and publications. There is an extensive network of federal information centers nationwide. Some have libraries; others have toll-free phone numbers, and some tie in to the nearest public library. (See the Appendix for a list of Federal Information Centers.)

For information on individual donors, your local public library is the best resource. Along with standard biographies, social registers, and club rosters, most public libraries maintain news clipping files on prominent local families and individuals.

With the advent of computer bases and microcollections, the art of exploiting the full resources of the present-day library has become more demanding. Many libraries have access to a computer. However, remember the computer is simply "a powerful helper,"[4(p.81)] and it is only as good as the information fed into it. Learn not to be intimidated by it. Computer files of possible interest to the grant seeker, to which your public or academic library may have access, are the Lockheed Dialog System (contains many files on the social sciences and some on grants), the FAPRS (a computerized file of the *Catalog of Federal Domestic Assistance*), and those compiled by Systems Development Corporation. The Foundation Center has several computerized files on foundations and their grants. To acquire access to these files, you (or your sponsor) must join the Center's Associates Program, which entails an annual fee plus additional charges for services performed upon request. For the most part, however, The Foundation Center's computer files are geared toward the needs of the organizational as opposed to the individual grant seeker.

CAN YOU (SHOULD YOU) HIRE SOMEONE ELSE TO DO IT FOR YOU?

The temptation is great to try to find someone to do the initial wading through information for you. Indeed, there exist various consultants, fund-raising firms, information services, and computer files to which you may subscribe that promise, for a fee, to perform your initial prospect research. However, even if you or your sponsor can afford it, I strongly recommend that you resist the temptation to let someone else do it for you, because no one else knows your project as well as you do. Some outsider, no matter how superficially familiar with your project, might miss important connections during the research phase that you, as the originator of the grant idea, would have picked up quite naturally. For this and other reasons, I concur with The Foundation Center's recommendation "that initially grant seekers invest their time rather than their money."[4(p.52)]

SOURCES AND TYPES OF INFORMATION AVAILABLE

For the four main categories of funders outlined in the previous chapter, there are various types of printed and other information available that are of varying degees of usefulness to the grant seeker. In this section, we will discuss the basic reference tools relating to each category of funder and some informational problems associated with these tools. Other publications of specific interest to the individual grant seeker will be discussed in the bibliography section, which follows this one.

FOUNDATIONS

Much of the information the grant seeker would like to have on foundations is hard to get. This includes requirements for eligibility, number of grants available, number of other applicants, percentage of proposals funded last year, your chances of securing an interview, the likelihood of your reviewing a successful application, the exact dates when the board meets, and when you will get a response. Answers to many of these questions are available for only a small number of foundations (400 to 500) that publish annual reports or issue newsletters, pamphlets, and other descriptive materials.

Yet of the four categories of funders, foundations are the most consistently documented, and information on them is the most centrally available. The following information is readily available on most of the 22,000 foundations: name, address, phone number, trustees, assets, and a relatively current (though not detailed) list of grants awarded. For all but the very largest foundations, little is available in terms of history or background, application procedures, funding patterns, or priorities. The grant seeker has to make use of the sketchy financial data provided on IRS returns, compare several years' grant listings, and draw his own conclusions.

The bible for the individual grant seeker is a relatively new publication: *Foundation Grants to Individuals* (New York: The Foundation Center, see latest edition). This directory includes information on those 1,000 or so foundations that make grants directly to individuals, the largest number being for educational purposes. To qualify for inclusion in this directory, foundations must make grant awards of at least $2,000, select their own grant recipients, and accept direct applications from individuals. The third edition (1982) is divided into the following sections: scholarships and loans, fellowships (research or work toward doctorates), grants for foreign individuals, internships and residencies, general welfare and medical assistance, and awards, prizes, and grants through nomination. (*Note:* In many cases, individuals can seek nomination or gain institutional support via a sponsor.) Grants restricted to company employees are also listed. In addition to various indexes, an excellent bibliographic essay by Jane C. Malik is provided. Entries in this directory are comprehensive, including much of the information the grant seeker needs to decide whether to proceed further in researching a given foundation.

There are two types of information available on foundations you may identify as potential funders via this directory: primary information (from the foundations themselves) and secondary information (from directories, indexes, and computer files compiled by The Foundation Center and other agencies and publishers as well). Always refer to additional information (primary, if possible) before mailing off a proposal to a foundation you've selected using *Foundation Grants to Individuals.*

If the foundation you're interested in publishes an annual report (less than 2% do), write and ask for it. The foundation annual report is the best single source of detailed and comprehensive data on grants, future commitments, officers and staff, application procedures, preferred formats, and deadlines. Much of this information is available nowhere else. Also request pamphlets, newsletters, or brochures, if available. Ask to get on the foundation's mailing list. Such primary information sources describe new program interests, recent grants, staff changes, proposal instructions, and occasional biographical information on staff members and trustees.

For the majority of foundations for which no other information is available (most small, family, or locally oriented foundations fall in this category), you will have to refer to the information returns (forms 990) that foundations file with the IRS each year. These primary research documents include basic financial data, names and addresses of officers (but not necessarily staff), and a complete list of grants (includes recipients' names and amounts but ordinarily contains little or no additional descriptive data). It isn't much to go on, but for most foundations it's all the grant seeker has. You may examine microfilm copies of the 990 forms at The Foundation Center libraries or cooperating collections, or you may request photocopies from the IRS. Keep in mind that 990 forms are usually a year or two out-of-date, because they refer to the previous fiscal year, and it takes time for IRS to process and photocopy them.

Secondary sources of information, which are somewhat less useful to the individual grant seeker, are:

The Foundation Directory. New York: The Foundation Center, see latest edition. This biennial publication includes approximately 3,000 foundations arranged by state with assets of over $1 million and/or annual grants in excess of $100,000. It has several indexes, the most heavily used (and misused) of which is "Fields of Interest Index." Keep in mind that this index refers to stated program interests, not actual grants awarded, and it covers national and regional foundations only.

Sourcebook Profiles. New York: The Foundation Center, see latest edition. This loose-leaf service analyzes the top 2,000 foundations in great detail. It is helpful if your potential funder falls in this top group. The main advantage is convenience of format, since most of the information included is available elsewhere.

National Databank. A comprehensive listing of all 22,000 foundations containing purely financial data, which is usually several years old. Arranged by state and within state by descending total grants amount, this listing should be used as a jumping-off point to identify small foundations in your locale and as an index to the IRS returns.

Foundation Reporter. Washington D.C.: J. Richard Taft Corporation, see latest edition. This covers much of the same ground as *Sourcebook Profiles,* described above.

Trustees of Wealth. Washington, D.C.: J. Richard Taft Corporation, see latest edition. This reference tool contains biographies on foundation officers and trustees. It includes some personal information not readily available elsewhere.

The Foundation Grants Index. New York: The Foundation Center, bimonthly in *Foundation News,* annual cumulated volumes, and microfiche cards containing "COMSEARCH Printouts" on approximately 70 subject fields plus geographic listings. This reference tool includes larger grants (over $5,000) of the top 100 foundations plus those of several hundred additional foundations that report to The Foundation Center. It has detailed indexing and provides an overview of the major foundations active in your field. Use it for referrals, suggested further research, and names of contact persons. The *Grants Index* is a good jumping-off point if your idea falls within an easily identifiable subject category. The editorial criteria for the *Grants Index* exclude grants to individuals, so you'll have to check further. The foundations whose grants to organizations are listed here may well make grants to individuals also.

Carol M. Kurzig, *Foundation Fundamentals.* New York: The Foundation Center, 1980. Although directed toward nonprofit organizations, this guide has many helpful hints on conducting foundation funding research. It includes sample searches, useful appendixes, and a good bibliography.

Foundation News, Washington, D.C.: Council on Foundations, bimonthly and *The Grantsmanship Center News,* Los Angeles: Grantsmanship Center, monthly. These are also directed to nonprofit organizations but occasionally have articles of interest to individual applicants. Other periodicals of specific interest to individual grant seekers are discussed in the bibliography section of this chapter.

There are many other directories and funding guides on foundations and their grants. Most cover similar territory and are not discussed here because they pertain to the larger foundations which are documented elsewhere.

CORPORATIONS

As mentioned in the previous chapter, corporations are relatively new as funders of major impact on the giving scene. Even now, most corporate grants are severely limited to specific categories of recipients (employees, relatives of employees, members of minority groups, and residents of certain locales). As indicated in Chapter 5, the grant seeker

must be extremely creative in finding ways to get corporations to fund
him. The same is true of finding information on corporate funders.

The reasons for the problems grant seekers encounter in getting
information on corporate funders are many: First and foremost, corporate grants decision makers fear being besieged by applicants. They are
also worried about muckraking journalists and the disapproval by stock-
holders of their grantmaking activities. The corporate funder's policy on
grants fluctuates from year to year depending on profits and the state of
the economy. In times of sharply shifting economic conditions, the desire
to maintain the utmost flexibility, coupled with executives' prerogative to
change their minds quickly regarding grant decisions, lead to hazy or
even nonexistent contributions policies. In other words, some of the information you need is not available simply because it doesn't exist. There
are no printed guidelines. There is no stated policy. Corporate grants may
represent a mere postscript to the public relations budget. Information
on corporate funders, moreover, is difficult to acquire, because many corporations tend to be quite secretive and because it is not centralized in
any consistent way in readily accessible reference tools or in computer
files.

Unfortunately, there is no directory or index of corporate grants to
individuals. Those subject directories that do exist are directed to nonprofit organizations and tend to focus on the arts and on education to the
exclusion of other subject fields. Several organizations have been active,
however, in conducting research on corporate giving patterns. The most
prominent of these are The Conference Board, which conducts statistical
and management studies on corporate contributions, the Business Committee for the Arts, and the Council for Financial Aid to Education, all
of which have offices in New York. You might contact these organizations or refer to their current research reports or other publications for
information on national funding trends and priorities among corporate
funders. In addition, the Better Business Bureau has files on many corporations and reports on their grants. Local chambers of commerce com-
pile lists of corporate contributions in your own geographic area.

Once you identify potential corporate funders, the information the
grant seeker would most like to have includes local and headquarters
address and phone numbers, list of top management personnel and rela-

tionship to the company, key grants decision maker(s), net sales and income last year, date contributions budget is set (often in June or September for the next fiscal year), existence of corporate foundation or contributions program, major subsidiaries, divisions, or plants in your area. Is an annual report available? Are last year's gifts summarized anywhere in print? To secure most of this information, you must have access to a wide variety of reference tools.

Use your imagination to try to overcome obstacles to getting what you need to know about corporations. Also, remember the importance of *personal contact*. Whom do you know who can help you gather business intelligence? Local attorneys, journalists, professors, politicians, and even congressional staff members and assistants can be helpful in providing valuable insight into local corporate funding priorities. Whom do you know who serves on corporate boards or works for a corporation that you might wish to approach for a grant? Can they provide you with an entrée to an important grants decision maker?

In addition to personal contacts, consider the following as sources of information on corporations.

Resources on Corporations: Organizations and People

Local chambers of commerce and boards of trade have data on businesses and businessmen in your own locale.

News reporters and editors often have the "inside dope" on local citizens and their companies.

Economic development offices have directories of state manufacturers and plant locations.

The office of the attorney general or county clerk's office in your state may have records on corporate foundations (since they are tax exempt) and in some states, corporate annual reports.

The U.S. Bureau of Labor Statistics maintains a Public Disclosure Room in Washington, D.C., where information on company assets is available.

The offices of trade associations, other professional groups, and unions have informed opinions about companies, industry studies, membership rosters, and analyses of national trends in funding.

United Way and other organizations that conduct annual fund drives have lists of corporate funders that they may be willing to share with you, especially if your grant idea appears to complement (not compete with) their own projects.

In my own community, the Greenwich, Connecticut, Public Library has an excellent business reference collection. If you live nearby, materials are available for public use on the premises.

Printed References on Corporations and Their Grants

Human Resources Network, *Handbook of Corporate Social Responsibility: Profiles of Involvement*. Radnor, Pa.: Chilton Book Company, 1975 (or latest edition). Case studies of corporate charitable activities in various subject fields.

Corporate Foundation Directory. Washington, D.C.: Taft Corporation, 1978–1980 (or latest edition). Reports on 321 corporate foundations. Indexes by state, fields of interest, corporate operating locations, names, sponsoring companies, and types of grants.

Corporate Foundation Profiles. New York: The Foundation Center, 1980 (or latest edition). Detailed analyses of 215 corporate foundations with financial data on an additional 500 corporate foundations with assets over $1 million or annual giving of $100,000 or more. Indexes by subject, type of support, and geographic focus.

General Information on Corporations

Most of these are available for your use at business or research libraries:

Directory of Corporate Affiliations. Skokie, Ill.: National Register Publishing Company, 1981. Tells who owns what and helps identify corporate subsidiaries.

The Fortune Directory of the 500 Largest Industrial Corporations, *Fortune Magazine,* annual (usually in mid-May issue), or *Fortune Double 500 Directory,* Trenton, N.J. (latest edition). Ranks top corporations.

Standard & Poor's Register of Corporations, Directors and Executives. New York: Standard and Poor's Corporation, annual. Provides corporate addresses, telephone numbers, names, and brief biographies of executives.

Million Dollar Directory. New York: Dun & Bradstreet Corporation, annual, 3 vols. (covering corporations with sales over $1 million) and *Middle Market Directory* enable the grant seeker to locate corporations by name, geographic area, and product classification.

Moody's *Manuals* are arranged by industry type and provide extensive information on the history of corporations, financial data, and lengthy lists of corporate officers.

SIC Manual. Standard industrial code numbers indicate what type of business a particular company is engaged in.

For information on smaller corporations, see local reference guides like *Becker's Guide* to the Chicago area (ask your librarian about these). Also, see Washington Researchers' *Sources of State Information on Corporations.* For biographical data, don't overlook various who's whos, especially *Who's Who in Finance and Industry.* Above all, don't forget one of the best sources of current information on corporations, businessmen, and corporate giving trends: *The Wall Street Journal!* Get in the habit of reading it regularly. Spend a day at the library perusing magazines like *Business Week, Fortune, Dun's, Forbes,* and any local periodicals or trade association newsletters. See *Business Periodicals Index* for help locating these. You'll pick up useful leads this way and valuable insight into local business activities and current priorities in corporate management.

GOVERNMENT AGENCIES

Concerning government agencies, the grant seeker would most like to have the following information: Should you apply to a local or regional office? Who's in charge? What forms do you need to fill out? Is the program funded now or just slated for funding? How many applications will be accepted? What proposal review system will be used? What evaluation criteria will be applied? Should you expect an interview? When will the results be made available? How much money has been designated for your locale? If you are funded, should you expect an audit? If you are not funded, can you reapply in the next grant-making cycle? Much of this information on state, local, and federal funders is available to the grant seeker, but not in one centralized location, library, or printed publication.

The Information Problem

The main obstacle to finding the right government funding agency for you is informational. For state and local agencies, there is an acute lack of specific information you need to submit an adequate proposal. In many cases, guidelines simply do not exist. There are multiple and confusing regulations so that it is difficult to ferret out which programs might pertain to your grant idea or, indeed, which agency is responsible for administering which program. Since there is so little in print concerning state and local funding mechanisms, you must rely on word of mouth by means of the "invisible network" described in Chapter 4. Nowhere is the rumor mill more active than in regard to government funding.

Follow up programs you see mentioned in local newspapers. Cultivate contacts you might have in any of the agencies, especially in your mayor's and governor's offices. Since state and local funding administrators are most susceptible to political influence, it does help to know someone, if not actually to get a grant, at least to find out for which programs you may be eligible. Above all, use your ingenuity. In researching state and local funding agencies, the telephone will be your most useful instrument.

Concerning federal programs, you will encounter an informational problem that is just the reverse, for there is an overabundance of printed material concerning federal government grants, a veritable mountain of guidelines, announcements, rules, and regulations, many of which appear to contradict one another. Take a logical approach to these documents. Try to ascertain which printed information is the most current and follow the recommendations or instructions contained therein. Use the telephone as much as possible to verify the existence of programs and your own eligibility for them. Discard obsolete materials immediately so as not to clutter your files. Read and reread relevant bulletins until you are sure you understand them.

Rules for federal grant seeking are often spelled out in circulars distributed by the Office of Management and Budget. The grant seeker should peruse relevant uniform administrative standards set for all funding agencies. Since they tend to relate more to large institutional funders than to individuals, the development officer or administrator at your sponsoring agency (if you have one) should be made aware of them.

Because there are masses of printed circulars, eligibility rules, and descriptive data on federal government programs, one library cannot contain it all; one individual cannot digest it all. So be selective. Research in depth only those government funders that really seem to be likely prospects for your grant idea. Above all, persevere.

Most federal government publications are issued by the Superintendent of Documents and are available for purchase from the U.S. Government Printing Office (GPO). Grant seekers should get on the mailing list for *Selected U.S. Government Publications,* which contains a fascinating array of publications, some relating to government funding, all available at low cost. When ordering publications, always include the GPO stock number and a check (for the right amount) made out to the Superintendent of Documents. If you have a sponsoring organization, it may be possible to open a charge account with the GPO. If you need a document quickly, there are GPO bookstores in major cities throughout the country that carry selected government publications. If you are operating on a low budget, keep in mind that many government publications are available free of charge if you request them directly from the federal agency whose staff compiled them rather than from the GPO.

Remember that information on government funders is liable to become outdated almost as soon as it gets into print. Check the source of the information carefully. Verify the data you receive from government agencies, officials, congressmen, state and local community leaders, commercial information services, and printed publications and newspapers by finding a seconding source as described earlier in this chapter. Get on mailing lists for government agencies whose funding programs seem to relate to your grant idea.

The following are major publications and resources of use to the grant seeker conducting research on federal government funders:

Catalog of Federal Domestic Assistance. Washington, D.C.: Executive Office of the President, Office of Management and Budget, annual, loose-leaf. This is the bible of federal government funding. It is a "government wide compendium of federal programs and activities" directed to state and local governments, territories of the United States, counties, cities, municipalities, public and private institutions, profit and nonprofit organizations, special interest groups, and individuals. It includes any program that provides assistance or benefits the American public. In other words, its coverage is extremely broad.

For example, the 1980 *Catalog* contains descriptions of 1,123 programs administered by 57 federal agencies. It is divided into three basic sections: (1) index, includes Agency Program Index plus Functional Index Summary, Popular Name Index, Subject Index, Applicant Eligibility Index, and Deadline Index; (2) program descriptions, detailed entries arranged by program numbers; (3) appendixes. This reference tool is so voluminous and intimidating that the temptation is great to throw up your hands in dismay. It looks as if it will take up weeks of your research time. By its very nature, much of the information you'll get will be out-of-date. However, the serious grant seeker cannot afford to overlook this source. First read "How to Use This Catalog." Also, the editors of *Grantsmanship Center News* published an excellent article, "How to Use the Catalog of Federal Domestic Assistance," in the November/December 1979 issue. Reprints are available at a nominal cost.

I recommend the following research methodology for the individual grant seeker who has never before used the *Catalog*. After reading the introduction, turn to the Applicant Eligibility Index chart. Look for programs with an X in the "individuals" column. Jot down their numbers. Also check the detailed Subject Index for cross-references to programs you may have overlooked, jot down these numbers too. Then proceed directly to the Program Descriptions section. Look up each program by number in which you have an interest. Read the descriptions carefully, especially noting any restrictions, limitations, or eligibility requirements that tend to exclude you or your grant idea.

A typical entry in the *Catalog* includes the following data: number and program name; users and use restrictions; eligibility requirements; application and award process; assistance considerations, postassistance requirements (reports, audits, records); financial information; program accomplishments; regulations, guidelines, and literature; information contacts; related programs; examples of funded programs; criteria for selecting programs. In other words, entries include all of the relevant data the grant seeker needs to determine whether to conduct further research on a particular government agency as a potential funder.

Next contact the information person whose name is provided by telephone or by letter. Most experts agree that the best place to start a search for funds is with the local office of the government agency. Ask for current guidelines, eligibility requirements, and application forms. Find out about the deadline. See if staff members will let you know informally whether or not you or your idea qualify for funding.

If you follow this research method, the *Catalog* will become much less confusing. It is really just a starting point to indicate the wide variety of federal programs available to the grant seeker. As noted earlier, there is a computerized version of the *Catalog* called *FAPRS* (Federal Assistance Program Retrieval System). In each state there are designated access points (state, county, and regional offices, agencies, some libraries, and universities) where, for a fee, computer searches may be conducted.

Keep in mind that although the *Catalog* is about as comprehensive as any research tool could be expected to be, it does *not* include federal revenue sharing programs, automatic payment programs, foreign aid, or solicited contracts. In addition, certain grant programs are not included because they fall through administrative loopholes, e.g., they are funded by various agencies via discretionary accounts rather than via a budgeted authority and, hence, are not defined by the editors of the *Catalog* as official "programs." The moral is always to know your reference tool, especially its limitations and exclusions.

Besides the *Catalog,* the following resources and general reference tools on federal government grants might prove useful to the grant seeker:

U.S. Government Manual. Washington, D.C.: U.S. Government Printing Office, annual. *The* guidebook to the federal government, senior officials, agencies, and programs; includes organizational charts and supplements.

Commerce Business Daily. Washington, D.C.: U.S. Department of Commerce. "Washington's Shopping List"[6] details everything the federal government wishes to procure; heavily used by those seeking government contracts.

The Federal Register. Washington, D.C.: U.S. Government Printing Office, daily, Monday to Friday. The most up-to-date printed source on new and proposed agency rules and regulations, some of which may be relevant to grant seekers.

Agency telephone directories. To get these, contact the public information office or publications division of the federal agency in which you are interested; they are also available from the U.S. Government Printing Office. Organizational sections of these directories help you learn how departments function and how to locate specific individual staff members and decision makers within a given agency.

The National Science Foundation, National Institutes of Health, National Endowments, and other federal government funding agencies issue their own program announcements, guidelines, annual reports, and newsletters. Get on their mailing lists.

The office of your own Congressman. Don't overlook this valuable resource. Your Congressman, his aides, and other staff members can provide inside information on federal agencies and officials, referrals, introductions, and interpretations of guidelines. They may even give you tips on how to word your proposal. Don't be intimidated by your Congressional office. Remember: "As a taxpayer and constituent, you are entitled to be served."[7]

Sources on State and Local Funders

Although printed resources on local government funders are sparse, the following reference tools might be useful to the grant seeker:

Some state and local governments publish their own funding guides (modeled on the *Catalog of Federal Domestic Assistance*). Ask to see these at your state library or order them from the issuing agencies.

Federal Funding Guides for Local Governments. Washington, D.C.: U.S. Government Information Services, annual, plus *Local Government Funding Report,* a biweekly newsletter detailing current developments.

There are various handbooks of state information, like George A. Mitchell's *The New York Redbook,* which provide background information on state officials, agencies, and departments, similar to the *U.S. Government Manual.* Look for these at your local public library, or ask the librarian.

The State Budget. A copy is available at your state library. If you can wade through it, this document provides valuable statistical backup data for the "need" section of your proposal, as well as a good sense of your own state's funding priorities.

Your mayor's office and state congressional offices can provide insight into local funding requirements, referrals, introductions, and technical assistance, the same as your congressional office in Washington, D.C. Ask for this help, since it doesn't come automatically. Keep in mind that contacts the grants seeker makes at the local funding level may prove worth their weight in gold.

INDIVIDUALS

As indicated in the previous chapter, the individual donor is the most mysterious, least documented source of potential funding. Since there exists no directory of wealthy individuals who wish to give away their money, in most instances you will hear of an individual funder by word of mouth. Or he will approach you himself or via an intermediary.

If the name of a potential individual funder is given to you or if you are lucky enough to have one approach you, any of the following information might be helpful: residence address and phone number; primary business affiliation/title, company address, and phone; other business affiliations; clubs, organizations, board positions; approximate financial worth; educational background; personal data on spouse, children, military service; awards and publications; previous gifts made by the individual and other family members; religious/political affiliations, interests, and hobbies. Who originally recommended his name? Always state this recommendation first when contacting potential individual donors.

The "outside interests" of an individual donor should not be taken lightly. They often serve as strong indicators of areas in which he might be willing to give away money. This type of research into someone's personal life, private interests, and finances may make you feel as if you're intruding on his privacy. This brings us to an important point about prospect research: You want to secure every possible piece of information on a potential funder, since you can never tell when an item will come in handy, e.g., knowing what hobbies or sports an individual donor follows is an important conversational ice breaker if you are lucky enough to secure an interview. However, use that information judiciously and with discretion. Much of what you will find out about an individual's financial situation and personal life will serve as "deep background" information only and will never appear as part of your proposal or in written or oral communications with the funder. You should never let on to a potential individual donor that you've performed all this research on his background and interests. He might be offended.

The same resources (and discretion) you use to get information on individual donors may be employed also for researching grants decision makers at various institutional funders. Particularly if you secure an interview with a funder, it can't hurt to become familiar in advance with the personal and professional backgrounds of those who will meet with you.

You should develop information on individuals from publicly available sources, but amplify data you receive this way through conversations with anyone who might know the man or woman. Your public library will have press clippings files on prominent local citizens as well as a wide range of biographical directories. See the *Biographical Dictionaries Master Index* for references to 53 who's who-type publications on living individuals. See also: *New York Times Index, Wall Street Journal Index,* and *Business Periodicals Index* for access to current print media coverage of individuals.

If you know an individual's occupation, you can proceed to specialized professional or vocational directories like *Martindale-Hubbel Law Directory, American Architects, American Men of Science, Medical Specialists Directory,* and *Standard & Poor's Register* for corporate executives. These kinds of directories vary in the amount of information they provide. Read the introduction to each publication to check the source of the data. Remember, what an individual says about himself in print may differ somewhat from the actual facts. *American Catholic Who's Who* is very useful for those individuals who qualify. For socially prominent individuals and families, see regional bluebooks and social registers in major cities. These tend to provide personal but not professional data on individuals.

Other resources include city directories available from realtors or from the publisher, R. L. Polk & Company. These directories give the basics: address, employer, phone number, and spouse's name. Even the telephone book can sometimes provide a clue to identifying potential individual funders. See the yellow pages under particular professions. Libraries and newspaper morgues have clippings files (many are now on microfilm), some of which contain helpful biographical and other data. Club membership rosters and donor lists of major nonprofit organizations (hospitals, colleges, museums) in your area provide insight into who has money to give away. Proxy statements for corporations registered with the Securities and Exchange Commission (SEC) provide extensive financial information about directors and their stock holdings. Copies are available from the SEC in Washington, D.C., and at SEC public reference rooms in major cities. For recently deceased individuals and information on bequests, estate probate files may be examined at the office of the county clerk or probate court in your locale.

In addition, the following reference tools may be useful to the grant seeker for research on potential individual funders:

Current Biography, New York: Wilson Company, monthly with annual cumulations. For individuals in the news.

Social Register. New York: Social Register Association, annual in November, updated by *Dilatory Domiciles* in January and *Summer Social Register* in June. Lists 20,000 of America's prominent families, grouped together with names and addresses recorded; gives clubs, organizations, and affiliations; indexes married and maiden names.

"The Richest People in America, The Forbes Four Hundred." *Forbes,* September 13, 1982, pp. 99–186. Brief biographical listings with some photographs of America's "centimillionaires"; includes notes about corporate and family affiliations, personal interests, and charitable activities; fascinating reading.

Trustees of Wealth. Washington, D.C.: Taft Corporation, 1979–1980 (or latest edition). Biographical directory of private and corporate foundation officers; geographic index.

Various who's who volumes mentioned earlier including *Who's Who of American Women, Who's Who in Advertising, Who's Who in Oil and Gas,* as well as regional who's whos.

ANNOTATED BIBLIOGRAPHY OF PRINTED SOURCES ON GRANTS TO INDIVIDUALS

CRITERIA FOR INCLUSION

In the previous sections of this chapter, I discussed the major reference tools for the individual seeking information on potential funding sources by category of funder. The following bibliography includes three additional groups of reference tools: general directories with overlapping coverage of several categories of funders, less important but still useful reference tools, and sources directed to specific target groups or covering particular subject fields.

The organization is simple: general reference tools are followed by more specialized ones. The first section of the annotated bibliography is arranged in alphabetical order, according to the name of the author, editor, compiler, issuing agency, or title, if none of the above pertain. The second section is arranged by population group, e.g., artists, social scientists, women, and so on. Various types of publications are covered, including directories, funding guides, bibliographies, periodicals, and checklists.

AN IMPORTANT NOTE FOR THE GRANT SEEKER

This bibliography is not comprehensive; rather, it is selective. It includes those reference tools that I believe to be valuable (some more than others) to the individual seeking grants. Each listing provides a basic bibliographic citation plus my own annotation, with hints on how best to use the reference tool and my evaluation of its usefulness to those seeking grants.

All publications listed here were examined by me. The publication date provided is the latest edition available at the time of my research for this chapter.

It should be noted that a printed bibliography of this sort by its very nature becomes rapidly obsolete. So *always* look for more current issues, revised editions, newly published directories, and updates in the libraries, publication lists, and information centers described elsewhere in this chapter. Check with your local public, academic, or special library to see if a specialized directory of funding opportunities exists in your field or in a related one. If you do find a given reference tool to be especially helpful, turn to the bibliography at the end of that particular publication. Such a bibliography may provide valuable leads to other resources not covered here. Good luck on this—perhaps most essential—part of your grant research.

Postscript: For the most part, directories geared exclusively to those seeking scholarships are not included here, since this book is not intended as a scholarship guide. However, directories that list apprenticeships, research grants, fellowships, and exchanges are listed. Please note also, although the prime criterion for inclusion in this bibliography is usefulness as a reference tool for the individual grant seeker, many of the reference tools covered also are valuable sources of information for those seeking funding for nonprofit organizations, especially for grant seekers with projects sponsored by large institutions.

GENERAL SOURCES ON GRANTS TO INDIVIDUALS

Annual Register of Grant Support (15th ed.) Chicago: Marquis Academic Media, 1981–1982, 781 p. A compendium of grant programs from a vast array of funders, including government agencies, foundations, associations, and corporations. Organized into eleven major categories with further subject and target group subdivisions, which make it relatively easy to zero in on the subject in which you have an interest. All of the 2,470 entries, including address, type of funder, purpose, eligibility, financial data, and application procedures, are in hard-to-read small print. This directory, heavily used by grant seekers, suffers from attempts at being too comprehensive. Its coverage includes all types of aid from all types of funders. Yet, the editors are vague in the introductory material about who the projected audience is. The programs listed are not limited to grants for individuals. Many are for the affiliated; you should double check eligibility requirements. Although this reference tool would be better if it were more selective and specific, it represents, nevertheless, a valiant effort to include everything and should not be overlooked by grant seekers, especially by scholars and scientific researchers.

ARIS Funding Messenger. San Francisco: Academic Research Information System, Pacific Medical Center, periodical, 1976–. *Medical Science Report,* monthly; *Social and Natural Sciences Report,* monthly (covers social, behavioral, and physical sciences, as well as engineering and business); *Creative Arts and Humanities Report,* nine times per year. A newsletter-type publication, the main purpose of which is probably to get grant seekers to order the application kits that the publisher distributes. Entries include name and type of funding agency, address, contact and phone, number of application kit, program titles, and description. There is also a short current awareness section on "meetings and deadlines." This periodical covers governmental and nongovernmental funders but provides especially good breakdowns on federal programs. The paragraphs on program description seem to me to be well written and worth perusing for most grant seekers, academics in particular. Very little is provided in the way of application procedures. (You must order the kits for that.)

Jan E. Balkin (Ed.), *1981 Federal Funding Guide.* Washington, D.C.: U.S. Government Information Services, 1980, loose-leaf. The brief introduction implies that this reference tool takes up where the *Catalog of Federal Domestic Assistance* leaves off. Although this may or may not be the case, it does cover some programs not listed in the *Catalog,* and its concise, readable format, especially the Quick Check section, makes it easier to use. Organized by broad subject categories and within categories by program, this tool describes over 140 grant programs. Since it is directed mainly to nonprofit organizations, grant seekers should refer to the "Are We Eligible?" sections of each entry to see if individuals may apply. There is no index.

Bank of America Corporation, *Bibliography of Corporate Social Responsibility Programs and Policies.* San Francisco: 1977, Vol. 6, 78 p. The editors of this bibliography have adopted a very broad definition of corporate social responsibility, which includes corporate grants. Organized into 20 key subject categories, this bibliography identifies articles, books, films, company reports, even speeches of possible interest to the grant seeker. Many of these, to my knowledge, are listed nowhere else. Useful as a guide for those seeking corporate funding as well as a rough indicator of which subject fields corporate funders tend to favor.

Barbara Coe (Ed.), *Cultural Directory II: Federal Funds and Services for the Arts and Humanities.* Washington, D.C.: Smithsonian Institution Press, 1980, 100+ p. This is an excellent reference tool for the grant seeker in almost any field of endeavor. Its title belies the contents, since it is not limited just to the arts nor just to grants. It covers more than 300 federal government programs to assist individuals, organizations and institutions with financial aid, employment opportunities, and the provision of services. Well researched and easy to use, what is unusual about this directory is that it lists government agencies seemingly unrelated to cultural concerns but with various programs to which grant seekers may apply, under certain conditions. It also suggests some fresh leads besides the National Endowments and other old standbys. Arranged by federal agency, each entry provides detailed listings, including "what/for whom" (see this to ascertain if individuals are eligible), program descriptions, and highly insightful comments about the original intentions of enabling legislation on various federal granting agencies. To use this tool efficiently, the grant seeker should look up specific subjects in the index and then refer back to the main text. Appendixes list various local, state, and regional offices of government funders. (*Note:* With this as with most secondary sources on government funders, you should double check, if possible, with the funding agencies themselves to be sure programs in which you have an interest are still in existence.)

William E. Coleman, *Grants in the Humanities.* New York: Neal

Schuman Publishers, 1980, 150 p. Directed to scholars, especially those conducting postdoctoral research in the humanities, the main text of this book bemoans the fact that most grants awarded each year go to scientists and social scientists rather than to humanists. However, the value of this book to the grant seeker lies in the appendixes. Appendix A, Granting Agencies, lists 130 funders that provide grant support to individuals for research and study. Entries are quite detailed, including how many, how much, conditions (e.g., eligibility), contacts, and application forms. This appendix also contains an excellent keyword-type index. Appendix B, Calendar of Deadlines, for programs listed in Appendix A, is arranged chronologically by week and month. Appendix E lists state committees that fund local humanities projects. This reference tool is quite useful, if you're a humanist.

Directory of Publishing Opportunities (3rd ed.). Chicago: Marquis Academic Media, 1975, 850 p. This reference tool lists 2,600 specialized and professional journals offering publishing opportunities for writers. Entries are arranged by 69 fields of interest, with full citations. Although not strictly about grant information, this directory is included here because, for many writers, the opportunity to publish is all the assistance they really require.

Denise S. Akey (Ed.) *Encyclopedia of Associations* (16th ed.). Detroit: Gale Research Company, 1982. Vol. 1, *National Organizations of the U.S.;* Vol. 2, *Geographic and Executive Index;* Vol. 3, *New Associations and Projects—Periodical Supplement.* This valuable reference tool has already been discussed in Chapter 2 for its "switchboard" function and its detailed descriptions of over 15,000 (!) national trade, professional, and other organizations. The grant seeker should use it for referrals to associations and professional societies in his subject field and especially as a source of information on potential sponsors. Many of the associations listed also make small grants, bestow awards, give scholarships, certificates, medals, and prizes, and sponsor contests, all of possible benefit to the individual grant seeker. (See Chapter 5 under Miscellaneous Category of Funders.) Arranged by broad subject category, it is a bit overwhelming, since there are so many entries. However, this reference tool is well worth the time it takes to peruse. Entries include association name (with key word in bold face), address, telephone, contact (chief official), founder, date, members and staff, state/local groups, description of activities (grants and awards programs are noted here), publications, meetings, and affiliated organizations. Use the alphabetical and key word indexes to refer to appropriate sections, if the subject arrangement fails you.

Fulbright and Other Grants for Graduate Study Abroad 1983–84. New York: Institute for International Education, 1982, 3 p. This compilation of grants offered by corporations, foreign governments, universities, and private donors technically is directed to graduate students but is useful to anyone seeking Fulbright or similar types of fellowships. Details and application procedures for other fellowships administered by IIE (one for each of 33 countries and a few teaching assistantships and creative and performing arts fellowships) are described. A word of warning to the grant seeker: the future of the Fulbright fellowship program presently is in doubt.

Alan E. Gadney (Comp.), *Gadney's Guide to 1800 International Contests, Festivals, & Grants in Film and Video, Photography, TV-Radio Broadcasting, Writing, Poetry and Playwriting, and Journalism* (Rev. ed.). Glendale, Calif.: Festival Publications, 1980, 610 p. An excellent reference tool for those wishing to enter their work in international contests, festivals, competitions, salons, exhibitions, markets, trade fairs, and other awards and sales events. (If you can't get a grant, you should consider these options.) Some of the agencies listed also provide grants, loans, scholarships, fellowships, residencies, apprenticeships, internships, training, and other personal benefit programs. Arranged by six media divisions and 166 subcategories, detailed entries include descriptions, number of awards, duration, amount, eligibility, judging, and deadlines. There are alphabetical and categorical indexes.

Frea E. Sladek and Eugene L. Stein (Eds.), *Grants Magazine, The Journal of Sponsored Research and Other Programs.* New York: Plenum Publishing Corporation, periodical, 1978–. Conceived as a forum to facilitate communication between grant seekers and funders, this periodical has articles on government, foundation, and corporate grants, especially as to technical aspects of the funding of scientific research. This is directed mainly to nonprofit agencies and research institutions, but it is also useful to the grant seeker in the sciences who already has a sponsor.

The Program Alert section lists currently available grant funds from a wide range of funders.

Howard Hillman, with Kathryn Natale, *The Art of Winning Government Grants*. New York: Vanguard Press, 1979, 245 p. Although more like a guide than a directory, this source, nevertheless, contains sections with valuable information for government grant seekers. Part II, Where the Money Is, describes how major federal agencies, quasi-government agencies, and some state and local funders operate, with addresses, divisions, and subdivisions. Part III, Information Sources, lists and evaluates publications, services, information centers, and consulting firms. It also contains an excellent selective bibliography. Although much of the specific information on government funders is becoming rapidly obsolete, this guide still provides interesting insights for the novice grant seeker into the government funding process.

Craig Alan Lerner, *The Grants Register 1981–83*. New York: St. Martins Press, 1980, 782 p. This is an international directory, mainly for those seeking grants for postgraduate studies or for other academics and scholars. It includes exchange programs, vacation and travel grants, as well as competitions, prizes, and honoraria for citizens of English-speaking countries. Although a bit difficult to use because of the tiny print, the subject index merits a careful scanning. Be sure to limit yourself to the United States or to countries in which you have an interest. Each entry includes purpose, subjects, number offered, value (amount), eligibility, closing dates, and contacts for further information.

David Magee, *How and Where to Get Capital: Dollars in Your Future*. Korville, Tex.: National Counselor Reports, Inc., 1975, 146 p. An unusual little book, it is included here for its upbeat approach to getting funding for grant seekers whose ideas qualify as businesses. The author's stated intention is to help the reader put his proposal into the hands of those seeking to lend, invest in, or help him with his plans (generally to the tune of $10,000 and up). It includes sections on sources of capital financing and loans, getting capital from the federal government, how to free yourself from debt, and a section on business aids, plans, and services.

Daniel Millsaps (Comp.) and the Editors of Washington International Arts Letter, *National Directory of Grants and Aid to Individuals in the Arts, International* (4th ed.). Washington, D.C.: 1981, 231 p. The Washington International Arts Letter started in 1962 as a citizens lobbying vehicle for the arts. This guide was initially intended as an indictment of U.S. funders for not adequately supporting artists. Strangely, the editors suggest that by its very paucity the guide serves its purpose. Unfortunately, this is one of the most frequently used reference tools by individual grant seekers, and it is disappointing. The main section is arranged alphabetically by organization. "Organizations," i.e., funders, include foundations, associations, museums, colleges, publishers, institutes, and other miscellaneous agencies. Very brief entries contain little more than organization name and address. No deadlines and no contact names are provided. In the left margin are code letters referring to disciplines, e.g., "A" for architecture, "C" for crafts, etc. This forces the user to leaf through the entire volume looking for his code letter or letters. The editors make a further distinction—upper case code letters for "professional" work and lower case code letters for educational support, e.g., "M" for music, professional work, and "m" for music scholarships or fellowships. This distinction seems to me to produce unnecessary confusion. The editors acknowledge that previous editions contained many errors. Check carefully (in another source if possible) any information you receive from this book.

Rita K. Roosevelt, Anita M. Granoff, and Karen P. K. Kennedy (Eds.), *Money Business: Grants and Awards for Creative Artists* (Rev. ed.). Boston: The Artists Foundation, 1982. 1 Vol., unpaged. Intended as a comprehensive guide to funding sources for individual creative artists (including poets, fiction writers, playwrights, filmmakers, video artists, choreographers, painters, printmakers, sculptors, craftsmen, and photographers), this book suffers from a lack of more detailed introductory material on what kinds of funders and programs are included or left out. The following categories are used: general listing, writers, visual artists, composers and musicians, state arts agencies, NEA. In addition, the index is cross-referenced by more specific artistic disciplines, e.g., architecture, crafts, drawing, and so on. Since the volume is unpaged, you must use the index under the abbreviations for the artistic disciplines in which you have an interest. This is a bit tricky until you get the hang of it. There are 278 entries. Each entry, numbered sequentially in chart format, has name, address, phone number, director or contact person, name(s) of program, eligibility (well researched), award, application, and deadline information. Since many of the grant programs listed here are limited to residents of particular states, the grant seeker should note carefully the eligibility requirements as provided.

A Selected List of Fellowship Opportunities and Aids to Advanced Education for U.S. Citizens and Foreign Nationals. Washington, D.C.: National Science Foundation, Spring 1980, 60 p. This brief listing appears to include government and association fellowships for international postdoctoral study. It has no subject index. This reference tool badly needs an introduction and/or an index. Who the compilers are or what are their intentions is impossible to determine. This list is worth a quick perusal, nevertheless, if your field happens to be international studies.

Social Science Research Council, *Fellowships and Grants for Research to Be Offered in 1980–81.* New York: 30 p. This is a pamphlet issued by the Social Science Research Council describing its own support of individuals through postdoctoral research grants. It includes, however, a brief section (pp. 6–10) on Fellowships and Grants of Other Organizations, which lists approximately 25 associations, foundations, and organizations which award fellowships and grants to individual scholars in the social sciences. Short entries provide little description but do give application dates and addresses to write for further information.

Washington Researchers, *Sources of State Information on Corporations.* Washington, D.C.: 1981, 50 p. Washington Researchers is an information-gathering service specializing in such subjects as markets, competition, economic conditions, and legislation pertaining to corporations. This guide was conceived because most researchers rely on federal information sources and miss out on the vast amount of data contained in state departments and agencies. This reference tool identifies various state offices that house documents describing local companies. Arranged alphabetically by state and within state by three subcategories, addresses, phone numbers, photocopying charges, and an occasional notation as to type of information are provided. Grant seekers should use this guide when they are stymied as to where else to look for public information on corporate funders or to help determine corporate assets.

Paul Wasserman, *Awards, Honors and Prizes* (5th ed.). Detroit, Mich.: Gale Research Company, 1981, 600 p. This directory lists over 4,000 different prizes and awards in a wide range of subject fields. The main text is arranged in alphabetical order by organization name. There is also a list of subjects and cross-references as well as a subject index. At first glance, this reference tool appears more valuable than it actually proves upon more careful study. Award seekers need to go further, since there is very little in the way of explanatory materials to help the user determine if he is eligible. Can you apply directly? Must you be nominated? No names of contact persons are provided to answer these questions. Grant seekers beware: as noted in the previous chapter, in some cases an award may mean a medal, citation, or a scroll rather than cash.

Virginia P. White, *Grants, How to Find Out about Them and What to Do Next*. New York: Plenum Press, 1975, 354 p. Intended primarily for the nonprofit agency, this easy-to-read guide contains advice that is useful to individual grant seekers as well as to those with institutional affiliation. The book is divided into two main sections: how to find out about grants (with special emphasis on federal government and foundation funders) and how to apply, with detailed instructions on writing proposals and constructing budgets. This is an excellent all-purpose guide to grantsmanship for the novice fund seeker.

Betty L. Wilson and William K. Wilson (Eds.), *Directory of Research Grants 1982*. Phoenix, Ariz.: Oryx Press, 1982, 398 p. This directory is geared toward academics in all fields, and especially to research administrators. That is to say, it is for the affiliated. It includes more than 2,000 programs offered by federal and state government agencies, foundations, associations, and corporations for research, training, and "innovative efforts." It is arranged by academic discipline and includes deadline dates. (The compilers claim that their information is at least 6 months ahead of the application deadlines.) *Note:* the compilers use the word *sponsor* where I would use *funder,* and this may be the source of some confusion for the grant seeker.

World Dictionary of Awards and Prizes. London: Europa Publications, 1979, 386 p. A dictionary of 2,000 international awards from 62 countries. For inclusion in this directory, the emphasis is on recognition of intellectual achievement. This reference tool lists prestigious lecture-

ships but excludes awards for sports, physical achievements, or fellowships. Each entry contains title, history, criteria, form of award (some
are cash but many are medals or certificates), how often made, name and
address of awarding body, and a list of current recipients. There are three
indexes. This is worth a glance if your grant idea is international in scope
or might attract foreign funders, for one reason or another.

SOURCES ON GRANTS TO INDIVIDUALS COVERING SPECIFIC SUBJECT
FIELDS OR DIRECTED TO PARTICULAR TARGET GROUPS

Artists

*Money for Artists, A Guide to Grants, Awards, Fellowships and Art-
ists-in-Residence Programs.* New York: Center for Arts Information,
1980, 4 p. This is an outline of government and private sources of funding
for individual artists in New York State. It contains details on the programs of the National Endowment, the New York State CAPS program,
and a two-page chart listing grants, lecture/demonstrations, workshops,
master classes, readings, and residence programs, with cross-references
by discipline for easy access. Few details or grant amounts are provided.
This is intended mainly as a starting point for artists to see what is available. The last page is an extremely useful bibliography of newsletters in
the arts that provide information on grants, awards, and fellowships. See
this bibliography to identify newsletters in your own field. For those
grant-seeking artists outside New York State, check to see if your own
state arts council distributes a similar outline.

*Ocular, The Directory of Information and Opportunities for the
Visual Arts.* Denver: Ocular Corporation, quarterly 1976–. This journal
contains articles of interest to visual artists. Each issue has sections on
grants, employment opportunities, competitions, exhibits, lectures, symposia, and workshops. The grants section includes about fifty programs
per issue, offering funding both for organizations and for individuals,
with and without deadlines. Arranged by state and within state alphabetically by funder's name, details on program, eligibility, amount, deadline dates, and application procedures are provided. Foreign funders are
included. Interesting graphics make this journal a pleasure to look at.

Robert A. Porter (Ed.), *Corporate Giving in the Arts 2*. New York: American Council for the Arts, 1981, 378 p. This directory contains information provided by the funders themselves on 502 of the country's largest corporations, accounting for $70 million in cash contributions. Intended for managers of arts organizations, this reference tool also is useful to the grant-seeking artist as a source of specific primary information on the corporations covered. Entries include address, contributions contact, phone number, total contributions, contributions to arts and culture, number of requests, kinds of organizations/activities supported, recent examples of recipients, geographic areas, evaluation criteria, annual reports, best time to apply, and requirements for application. Arranged alphabetically by company or corporate foundation name, the following indexes are included: Index by State, Index by Type of Arts Organization/Activity Supported, and Index by Type of Support Given. See this last index under the headings "Commissions to Artists," "Corporate Art Purchases," and "Special Projects" for help identifying potential funders of individual artists.

Virginia P. White, *Grants for the Arts*. New York: Plenum Press, 1980, 360 p. This is an excellent resource covering all aspects of finding out about and securing grants for projects in the arts. Directed mostly toward art organizations or artists with sponsors, this guide provides many insights into how grants are awarded and specific suggestions as to sources of further information. The section on government support of the arts, in particular, is quite useful and comprehensive. (As always, check with current sources to be sure programs listed are still extant.)

Film and Video

Film, Video Festivals and Awards, No. 3 (Rev. ed.). Washington, D.C.: American Film Institute, National Educational Services, 1980, 30 p. (FACTFILE series). This is a select list of U.S. and foreign film festivals and awards, both competitive and noncompetitive, some by invitation only. It includes a month-by-month chronology. Brief listings contain all relevant details, but this booklet is still difficult to use because festivals that provide cash prizes and awards and those that do not are comingled.

Steve Penny, *How to Get Grants to Make Films and Video, a Guide to Media Grants, Film, Video, Audio Visual Projects and Media Scholarships.* Santa Barbara, Calif.: Film Grants Research, 1978, 128 p. A lengthy introduction, based on the author's own experience seeking grants, includes many fund-raising hints. The remaining text contains a section on the Endowments and other federal agencies with detailed eligibility requirements and information contacts, a section on private foundations giving on the national level and grant programs on regional or special interest bases, and listings of academic scholarships, fellowships, and media research grants. This is a useful guide for the filmmaker, even though many of the programs listed require institutional sponsorship. See especially the appendix on budgeting.

Historians

Grants and Fellowships of Interest to Historians, 1981–82. Washington, D.C.: American Historical Association, 1981, 88 p. This guide contains over 180 entries for students and scholars. Funding sources include federal and state agencies, local organizations, foreign countries, foundations, professional societies, and awards committees. It is divided into three sections: dissertation and postdoctoral research, predissertation study and research, and support for organizations in historical research or education. It also includes a section on book awards and prizes. Entries are textual in nature and a bit difficult to use, but they do include names and phone numbers. There is no index.

Mathematicians

Fellowship and Research Opportunities in the Mathematical Sciences 1975–76. Washington, D.C.: Office of Mathematical Sciences, National Research Council, 43 p. Since this is the last edition of a discontinued annual publication, most of the specifics included are well out-of-date. This book contains two sections: information for predoctoral students and research opportunities for Ph.D.-level mathematicians. It includes fellowships, research grants, independent study, and employment opportunities and has an alphabetical index.

Minorities

See *Women.*

Musicians

Chaddie B. Kruger (Comp.), *Post-Baccalaureate Grants and Awards in Music.* Reston, Va.: Music Educators National Conference, 1974, 40 p. Although out of date, this booklet is useful because it is directed to mature musicians who may or may not have successful careers. It includes grants and awards in composition, conducting, instrumental and vocal performance, study abroad, and music research. It is divided into two sections: independent awards and graduate school programs. Alphabetical entries include brief descriptions and addresses but no contact persons. Use this as a jumping-off point for more current research, if you are a mature musician.

The Musician's Guide, the Directory of the World of Music (6th ed.). Chicago: Marquis Academic Media, 1980, 943 p. This vast compendium covering all aspects of the musician's world has three sections of possible interest to grant seekers: The National Federation of Music Clubs Scholarships and Awards Chart 1979–81 provides the name of the award, age limit, entry fee, amount (most are small), chairman, and address. The section on foundations includes approximately 150 funding organizations, but not all are foundations; some are art councils, clubs, commissions, and associations. Brief information as to what percentage of the funder's grants go for performing arts and eligibility requirements are provided. *Note:* Some of the funders listed do not give directly to individuals. The section on Music Competitions is a comprehensive alphabetical listing with brief descriptions. The musician seeking a grant should use this reference book as one of many resources on the subject.

Photographers

Jacob Deschin, Want a Photographic Grant? *35mm Photography.* Winter 1973, pp. 37–39; 112–116. Although obviously outdated as to specifics, this article is included here because it stresses the tough competition among grant seekers, provides many examples, and gives some insight to photographers (and other grant seekers as well) on the best approach to potential funders. This is not a directory listing but rather a textual discussion of funders in photography.

Linda Moser, *Grants in Photography—How to Get Them*. Garden City, N.Y.: American Photographic Book Publishing Company, 1978, 88 p. In the first half of this excellent guide, the author strongly recommends that individual grant seekers band together and incorporate as tax exempt organizations, find sponsors, or become employees of appropriate institutions in order to create funding opportunities where none seem to exist. The second half is a survey of those grant programs that do exist for individual photographers, including descriptions of about a dozen funders, case studies, details on the programs of the National Endowment, a helpful list of organizations, and a bibliography of other publications.

Playwrights

Michael Earley (Ed.), *Information for Playwrights*. New York: Theatre Communications Group, 1979, 22 p. This is an informative little booklet, the purpose of which is to outline the broad range of resources available for playwrights. The largest portion of the booklet is an annotated listing of nonprofit theaters seeking to produce new plays. The small section devoted to fellowships, grants, awards, loans, and competitions is well researched. It includes current information, names of contact persons, deadlines, and application data. This is a useful reference tool, well worth a careful perusal if you are a playwright.

Poets and Writers

CODA: Poets and Writers Newsletter. New York: Poets and Writers, Inc., five times a year, 1973–. This is a helpful, informative, and well-written newsletter that serves to update the *Poets and Writers' Awards List* (see below) and includes many short articles of interest to fiction writers. It is full of useful information for writers to keep abreast of upcoming courses, conferences, residencies, research, and employment opportunities. The Grants and Awards section lists recent winners of poetry and fiction prizes. The Deadlines section tells when to apply for upcoming grants and provides brief descriptions.

Grants and Awards Available to American Writers (12th ed.). New York: P.E.N. American Center, 1982, 90 p. This much-used directory details domestic and foreign grants of $500 or more for all types of writers. It is updated annually. The main body is an alphabetical listing by funder (foreign countries included) that provides name and address, title and amount of grant, deadline, eligibility, and to whom one should apply. There is also a listing of state arts councils with phone numbers, an Index of Awards, and an Index of Organizations. It includes grants for which an individual writer may apply directly as well as those for which a publisher or academic institution must make the nomination. P.E.N. American Center publishes a similar list for foreign writers.

Literary and Library Prizes (10th ed.). New York: R. R. Bowker Company, 1980, 651 p. This directory includes international, American, British, and Canadian prizes. Brief descriptions of each prize indicate where to write, the names of recent recipients, and the titles of their publications. There is an index of persons and awards but no key word or subject indexing. Remember, some prizes are cash, others are plaques, medals, or certificates.

Poets and Writers Awards List. New York: Poets and Writers, Inc., 1978, 21 p. This is an excellent listing of 100 grants, fellowships, and prizes offered in the United States to poets and fiction writers. Those grants listed provide a minimum of $500 in cash or publication awards. Detailed entries include phone numbers, announcement dates, deadlines, and names of recent recipients. Programs are for poetry, fiction, and translation. There is even a section on emergency money, listing four sources. This list is updated in *CODA, Poets & Writers Newsletter,* which includes deadlines, judges, and winners.

Poets and Writers Sponsors List. New York: 1979. This list includes over 600 national organizations that sponsor programs involving fiction writers and poets. Intended as a source for the sponsors themselves to make referrals, grant seekers can use this book to identify organizations that hire poets and writers for a fee to conduct readings, lectures, and other writing-related training programs. Sponsors include schools, colleges, museums, arts councils, libraries, bookstores, Y's, community centers, and religious groups. Arranged by state, each entry provides name, contact person, address, phone number, number of writers sponsored, and a brief description. The actual amount of the fee is not listed, but there is an occasional reference to the size of the sponsor's budget.

Scholars

Aids to Individual Scholars, Competitions to Be Held in 1982–83. New York: American Council of Learned Societies, 1982, 19 p. For grant seekers who already have their doctorates or equivalent backgrounds, this guide lists grants primarily for research. It includes research fellowships (to a maximum of $15,000 each), language training in Chinese and Eastern European studies, grants-in-aid ($3,100 maximum) for "special programs of humanistic research," travel grants for humanists to attend international meetings abroad, and exchange programs with the USSR. The guide contains detailed eligibility requirements and stipulations, which the grant seeker should read carefully.

Social Scientists

Kathleen Bond (Comp.), *Federal Funding Programs for Social Scientists.* Washington, D.C.: American Sociological Association, 1980, 84 p. This slim guide describes over fifty federal programs that support social science research, while noting that most federal programs are "mission" rather than research oriented. It is arranged by cabinet level department and independent agencies with relevant subdivisions. There is no index, so the user must thumb through the entire text. Many of the specifics provided here are somewhat out of date. But this guide is still useful for the detailed entries with examples and the fact that it includes the programs of a number of agencies that one might not think of as strictly involved in the social sciences. The grant seeker in the social sciences might well leaf through this guide as a source of ideas for new approaches to funders.

Women

Financial Aid: A Partial List of Resources for Women. Washington, D.C.: Association of American Colleges, Project on the Status and Education of Women, 1979, 8 p. This booklet is geared mostly to undergraduates, but it also describes programs for older women returning to school, grants for professional and technical training, and sources for loan funds. It lists state and federal programs as well as schools, foundations, local businesses, professional associations, and civic groups, and it contains a selected bibliography. This booklet could be dubbed a "mixed bag" of resources for women.

Marilyn W. Richards, *Funding Resources for Women in Development Programs*. Washington, D.C.: New Trans Century Foundation, 1978, 96 p. Under a grant from AID, the New Trans Century Foundation provides management and program development services to private and volunteer organizations. Much of the information included in this guide is applicable to those individuals affiliated with sponsoring organizations. Sections include U.S. private foundations, government agencies, U.N. agencies, U.S. private volunteer organizations, and multinational corporations. The latter section is especially helpful for its discussion of information sources and fund-raising approaches to profit-making corporations. Entries include key people, interests, geography, types of grants, policies, funding level, and sample grants. This is worth a look if your grant idea falls in this area.

Linda Byrd Johnson and Carol J. Smith (Comps.), *Selected List of Postsecondary Education Opportunities for Minorities and Women*. Washington, D.C.: U.S. Government Printing Office, 1981. 1 vol., various pagination. For "truly needy" individuals wishing to obtain financial aid for education or to pursue career goals, information is provided on loans, scholarships, and fellowship opportunities from private and public organizations/associations as well as from state governments. Sections include general sources, opportunities in selected fields of study, e.g., architecture, engineering, and so on, opportunities exclusively for women, American Indians, with the U.S. military, and five federal financial aid programs. Entries are lengthy and in paragraph format. No index is provided.

COMPENDIA IN YOUR SUBJECT FIELD

I have already mentioned journals and newsletters in your field as valuable current awareness tools and sources of information on upcoming grant programs. One final resource not to be overlooked consists of compendia to various fields and professions. Many have brief sections on funding opportunities. Most are published annually. The following is a partial listing of some compendia and directories that, depending upon your field, might contain information useful to your grant-seeking effort:

American Art Directory
Artist's Market
Audiovisual Marketplace
Craftworker's Market
Conservation Directory
Dance Magazine Annual
Directory of Religious Organizations
Fine Arts Marketplace
International Directory of Theatre, Dance and Folklore Festivals
National Directory of Mental Health
National Directory of Performing Arts
Official Museum Directory
Photography Marketplace
Research Centers Directory
Songwriter's Market
Writer's Market

Look for the above at your local public, academic, or special library. They will prove useful throughout your grant-seeking endeavor.

NOTES

[1] Kohl, K. A., & Kohl, I. C. *Financing college education.* New York: Harper & Row, 1980. P. 51.

[2] Broce, T. E. *Fund raising: The guide to raising money from private sources.* Norman: Univ. of Oklahoma Press, 1979. P. 169.

[3] Kurzig, C. M. *Foundation fundamentals*. New York: The Foundation Center, 1980.

[4] Barzun, J., & Graff, H. F. *The modern researcher* (Rev. ed.) New York: Harcourt, Brace, 1970.

[5] How to use the revised *Catalog of Federal Domestic Assistance*. *Grantsmanship Center News*, November/December 1979, pp. 23–37.

[6] Smith, C. W., & Skjei, E. W. *Getting grants*. New York: Harper & Row, 1980. P. 116.

[7] *Grantsmanship Center News*, January/February 1980, P. 37.

" . . . don't assume that an outstanding proposal is the key to funding. Its importance depends on the funding source. But, no matter to whom you're talking, it's always best to have a solid plan on which to base any discussion of funding."

—NORTON J. KIRITZ
Program Planning and Proposal Writing

Chapter 7 / THE PROPOSAL

As I stated at the outset of this book, far too much attention is, in my opinion, paid to the proposal. Since the most valuable use of any grant seeker's time is finding the right funding source, I would like to suggest that you view the proposal as a natural outgrowth of the research you have done identifying potential funders and learning as much as possible about funders' grant-making policies and preferred procedures for application. If you have done this homework thoroughly, your proposal will seem to write itself.

Some funding agencies, however, provide very little in the way of concrete guidelines to follow in constructing an adequate proposal. For those individuals who have never written a proposal, the prospect of doing so may seem highly intimidating. The purpose of this chapter, therefore, is to provide some clues to help you write a successful proposal within the framework of the guidelines of the funding agency itself. If there are no guidelines, this chapter sets forth a general outline you can follow that will give you the confidence to go ahead and write a proposal.

Since there are exceptions to every rule, the generalizations I make here are just that. Much of the information provided, nonetheless, will pertain to most instances of proposal writing. On the other hand, the individual grant seeker should keep in mind that there may be different kinds of formats to follow for different types of grants.

The complexity of the proposal you eventually submit may depend on the amount of money you are requesting. For an award in recognition of achievement, a short letter might do; for a fellowship, filling out a form might suffice; for a research grant, a more detailed proposal might be required. The category of funder and the review procedures your funder ordinarily follows will have a further impact on the proposal you submit. For example, corporations tend to be more flexible about the actual proposal format than foundations or government agencies; but they also tend to make smaller grants. (See Chapter 5.)

If you are applying on your own, you may require a slightly different type of proposal than if you are applying under the auspices of an umbrella group, of if a sponsoring organization is applying·to a funder on your behalf. This chapter describes in detail *all* of the elements that make up a full proposal. If you are only writing a short one, you may wish to select only those elements that will enhance your presentation to a funder. It's important, neverthelss, that you understand all these elements in case you are called upon to submit a longer proposal at a later date. If you are presently unaffiliated, once you have built up your reputation through the receipt of various awards, honors, and small grants you may wish to "trade up" to a larger grant, requiring institutional sponsorship. At this point, you (and your sponsor) will need to call upon the information contained in this chapter.

When you adopt the rules and recommendations provided here to your written proposal, common sense is a must. Remember, the purpose of your proposal is to present you and your ideas in the most positive way to a potential funder. If following a slightly different format or emphasizing one aspect of the proposal package seems to bolster your grant request, by all means do it. In other words, remain flexible.

The important thing to remember about the proposal is that it is (or should be) custom designed to present your own ideas to one particular funder. If the funding agency provides you with instructions as to the length, format, and contents, you should follow them—to the letter. Any advice I give you regarding the general format of a successful proposal should be disregarded if the funder you have in mind has its own preferred format. If a funding agency has its own policies on the contents of the proposal, their requirements automatically should supersede my advice.

Some funders have printed guidelines and instructions; others use application forms in which you simply fill in the blanks. A number of the suggestions you will find in this chapter on the construction of successful free-form proposals could be applied to filling out printed application forms as well. Many of the latter are in roughly the same format as the one I recommend: a statement of need, objectives, procedures, and evaluation. When filling out forms, as in writing proposals, remember always to follow the funder's instructions.

One final word of advice: when you sit down to write the proposal or to fill in the application form, try to relax. If you have a sound idea and you've done your homework thoroughly, you're already on the road to receiving a grant. There is really very little your proposal can do to jeopardize your chances at this point—unless you are outright careless. Follow these simple rules and let your proposal flow from your pen.

PRELIMINARY LETTERS OF INQUIRY

It is only fair to point out that there are two schools of thought on whether you need a full proposal when you first apply to most funders. In these days of increased competition for fewer grant dollars, grants decision makers may feel particularly harassed because of the flood of applicants. Yet there is no evidence to suggest that funders are responding by increasing the number of staff members to process proposals. The reverse, in fact, may be the case. A well-conceived preliminary letter of inquiry, therefore, is frequently preferred *by funders* to full-blown five- or ten-page proposals.

A letter of inquiry usually is one or two pages long. It briefly states why you are contacting a particular funder (e.g., John Jones of such-and-such an organization said you should, or because of the funder's demonstrated interest in your field). It sets down your credentials, your idea, and how much you need. It asks funding executives to let you know if they have an interest either in meeting with you to discuss the matter further or in having you submit a formal proposal.

Those who favor the use of the preliminary letter of inquiry point to the desire to save time and effort on the parts of both grant seekers and grant makers. They insist that it simply takes too much time to submit a proposal without some minimum guarantee of interest on the part of the funder. Furthermore, the preliminary letter of inquiry increases your chances of a quick response and allows you some room to modify your idea or procedures if the funder is interested but wants certain changes. Proponents of this method suggest that sending out initial feelers is good psychology, somehow permitting funding executives to believe that your proposal is their idea. Sending out a preliminary letter might even preclude your submitting an unsolicited proposal since, after your initial inquiry, the funder may invite you to apply in a more formal way.

Although all of these arguments might be made for those grant seekers applying to staffed foundations and to other large funders, sending out a preliminary letter of inquiry is a fairly dangerous procedure for the grant seeker applying to smaller, unstaffed or understaffed funders who might then seize the opportunity to turn you down without further ado. In other words, you might never get a chance to sumit a full proposal.

For the individual grant applicant, your preliminary letter of inquiry may be your *only* real contact with the funder—your one shot at a grant. So if you choose to take this route, you had better make sure that your letter is a good one. Besides, unless you are famous in your field or associated with a highly prestigious institution, it is unlikely that funders, after reading your letter, will invite you to apply. For these reasons and others, I personally prefer the construction of traditional, more formal proposals for most grant ideas.

The proposal is an important planning tool for your grant project. Writing a proposal forces you to get your ideas down on paper and facilitates the entire grant-seeking process. Once you actually have a proposal in hand, the worst of it may be over for you. Applying for funding for the ideas presented may seem easy by comparison.

Since it is far simpler to condense a full-blown proposal into a shorter version later on than it is to flesh out a hazy concept paper into a formal proposal, it will help you to put together a full proposal *before* you contact the funder. You will probably have to write a complete proposal eventually, so why not do it now?

Even if the funder you have in mind states that preliminary letters of inquiry are preferred, you will find that having a full proposal on hand, ready to go, will give you confidence in the initial stages of applying for funding. That is why I feel that the preliminary letter of inquiry should serve as an adjunct to, but not a replacement for, the proposal.

PERSPECTIVES ON THE PROPOSAL

From my experiences conducting grantsmanship workshops, it has become evident to me that there are ways of viewing a proposal that make it easier to write one. It helps to consider your proposal as nothing more or less than a verbal presentation of your ideas. Now you may say that it is very difficult to put a complex idea on paper. Nevertheless, if you cannot do this effectively, your chances of being funded are practically nil. Although it is impossible for you to be truly objective about your idea, make every attempt to be as honest as possible about its chances of success. Don't make grandiose promises you won't be able to keep. One rule to keep in mind is to think of your proposal as a recipe or a set of instructions for someone else to follow. In that way, it will be clear and concise with one step logically leading to another.

Although a proposal is nothing more than a written expression of a viable solution to a problem, it may also be viewed as promotional material. Your idea may be worthwhile, even brilliant, but it must be presented in a way that is sufficiently attractive so the funder will buy it. Without making exaggerated claims, you should project results that will be desirable to the funder. Select language that allows your enthusiasm to show through and intimates that you *will* be successful.

Do not permit the requirements of the funding agency to shape your entire proposal. There is one view which holds that you should shop around for an idea that seems to be popular with funders and write that idea up into a proposal. Then once you receive the grant money, you can modify the project to do what you really wanted to do in the first place. I am opposed to this procedure both on moral and on practical grounds. In the first place, it is truly unethical. Second, I don't believe you can get away with it, certainly not more than once. Remember, grants decision makers frequently communicate with one another about potential grantees. If you do get a grant this way, you really are no closer to accomplishing your true goal, which is seeing your idea in action, not just getting yourself funded.

How then can your proposal help make your idea more attractive to a funder? A sound technique is to put yourself in the other fellow's shoes. Ask yourself this question: Why should they give me a grant? Whatever answer you come up with, stress that aspect of your idea in your proposal. Keep in mind that the priorities of the funding agency may be different from your own. Find a way to connect the two. Try to identify the self-interests of the funder, and see if they match your own in any way. Your proposal should describe how your idea will help fulfill the grant-making goals of the funding agency.

The successful proposal establishes a *link* between the grant seeker and the funder. Up until recently you may have been viewing potential donors as adversaries. Again, dispel this thought from your mind. Other considerations aside, funding executives probably would not be in the business unless they derived some personal satisfaction from the pure and simple act of giving. This fact should be a source of reassurance to the grant seeker. The funder's money and resources coupled with your talent and ideas enable you to work together to solve an important problem or even to make a meaningful contribution to society. Your proposal should suggest that you are a potential partner rather than just a supplicant or worse, an adversary trying to chip away at the funder's most valuable resource—money.

It is important that you not lose sight of the human element of the grant-making process. You probably regard yourself as a relatively resourceful but ultimately powerless individual faced with a monolithic structure, the grant-making bureaucracy. This is a destructive image to be banished from your mind. Remember, in no case will your completed proposal be processed entirely by computers. At some point, an individual, a human being like yourself, will sit down and read it. Picture that human being across the desk from you as you write. Address your proposal to this person. Keep in mind that you are both players on the same team. I stated earlier that your idea should be set forth objectively in factual terms. However, the human element should not be dismissed completely. In the final analysis, funders are supporting people and not just ideas.

Another encouraging aspect to consider, especially for those who have heard how difficult it is for the unaffiliated individual to get a grant, is that it is more difficult to turn down a real person than a relatively anonymous agency. Grants decision makers who work with individuals report that they derive pleasure from the person-to-person relationship, which is much more readily imagined and for which the need appears more vivid than the person-to-organization one. It is, therefore, advisable that you use your proposal (and other contacts with grants decision makers) to project an image of yourself as someone who should not be turned down.

Like many investors, grants decision makers hope that a small amount of money will bring about a great deal of change. Although some funders may be content to help a few isolated individuals solve their own financial problems to get on with their work, most funding executives prefer to take a broader view. They want to feel that their investment in you will have a far-reaching impact on society at large. Your idea, in other words, must be one that others will be able to share, copy, or benefit from in some way.

No funding agency wants you returning year after year asking for grant money. Bearing this in mind, when you write your proposal, remember to design a "cup, not a funnel."[1] Your proposal should be self-limiting with projected dates both for commencement and for completion. This does not mean that you may never go back to the same funder for renewal of a grant. It does mean that your proposal should describe certain concrete goals that you plan to accomplish within a specific time period (usually one year although some funders do give two-, three-, and even five-year fellowships and research grants). When that period is up, you may require additional funding to accomplish further objectives, enlarge upon what you've already successfully achieved, or direct your ideas toward a broader audience. You will probably then have to write up an entirely new proposal for this next stage of your work.

Some savvy individuals begin work on their next proposal soon after they receive funding on their initial one. This is reasonable if you consider the time involved in the grant-seeking process and your wish to avoid long delays in your work while you await notification regarding funding. However, there are two points of which you should not lose sight. One is that each proposal outlines specific work you will accomplish that should, if necessary, stand on its own without future funding. Second, your primary aim is to get on with your work, not simply to get funded. In other words, don't allow yourself to become so caught up in the proposal-writing process that it usurps valuable time that you would otherwise be spending on your work. Proposal writing is a means to an end and not an end in itself.

COMMON CHARACTERISTICS OF THE SUCCESSFUL PROPOSAL

Each proposal is a unique document designed with a specific funder in mind. There are, nevertheless, certain ingredients common to most successful proposals. A good proposal usually accomplishes all of the following:

- Delivers an important idea *and* addresses a significant problem of society at large
- Indicates that the applicant has chosen an outstanding approach to solve the problem and a reasonable plan for implementation

- Assures the funder that the applicant is capable of success
- Shows that the proposal is, indeed, within the scope of the funder's activities and will serve in an immediate way to advance the grant-making goals of the particular funder to which it is addressed
- Sets forth anticipated results that seem to justify the costs and the time estimated to accomplish the project

THE PROPOSAL PACKAGE

By the proposal package, I refer to the cover letter, title page, abstract, and appendix that accompany your proposal and make it easier for the funder to understand and to process. The proposal package reflects the quality of your ideas being presented to the funder. It also has some impact, albeit subliminal, on the impression you make as a grant applicant.

Avoid fancy bindings or other expensive trappings that go along with your proposal. Rather than helping to underscore your creativity as you might wish, these extras suggest to a funder that you are wasteful and ostentatious. Remember to keep it simple. If an element of your proposal package does not serve to clarify your ideas, throw it out. The less bulky your proposal appears, the better. Always follow available instructions from the funder on what should be included in your proposal package, and be careful not to omit required information.

COVER LETTER

The cover letter is more than just a letter of transmittal. It enables you to make the application process less anonymous and more personalized. Rarely, if ever, should a cover letter be addressed "Dear Sir" or "To Whom It May Concern." If your research has not turned up the name of a single executive connected with a potential funding agency, you probably should not be applying to that agency. You do not know enough yet about the funder to warrant submitting a proposal.

Application guidelines or other printed materials such as annual reports usually indicate to whom a proposal should be addressed. Always double check to make sure that the executive in question is still employed by the funding agency. Obviously, it will not help your chances to address your proposal to someone who is retired, is deceased, or has been fired.

In the case of foundations for which printed guidelines are unavailable, you should address your proposal to the name in italics in the entry for that foundation in *The Foundation Directory.* If no name is in italics or if the foundation is not listed in the *Directory,* address yourself to anyone with "Executive" as part of his or her title as it appears in a list of officers and directors in the foundation's current IRS form 990 (described in Chapter 6). If no one has the word "Executive" in his title, address yourself to the President, Secretary, Treasurer, or to the first trustee on the list in that order of priority. If the foundation in question is included in the publication, *Foundation Grants to Individuals,* the name of the person to contact ordinarily is provided.

In the case of corporations, it is sometimes more difficult to ascertain to whom your proposal should be addressed, especially if application guidelines are unavailable, as is quite often the case. The simplest thing to do is to call up corporate headquarters (fortunately phone numbers are generally available) and ask the receptionist for the name of the executive in charge of corporate contributions. If the corporation has no one with this function, ask who handles public relations. Just about every corporation has someone in this slot. That executive should be able either to help you himself or to refer you to someone who will. While you're on the phone, you might be tempted to make a quick informal presentation of your grant idea. If you decide to do this, be prepared to answer any and all questions the executive might ask. Also, be ready to cut it short (*before* he turns you down on the spot) if he seems uninterested.

The following are examples of the wide range of titles given to corporate grants decision makers. To determine who's in charge, look for executives with similar titles in corporate directories, annual reports, news clippings or other literature: Manager, Educational Relations and Support; Vice-President for Corporate Contributions; Executive Secretary of Contributions Committee; Director of Public Affairs; Vice-President (Manager, Coordinator) for Public Relations; Coordinator of Support Programs; Vice-President for Community Affairs; Urban Affairs Coordinator. If you are unable to discover the name of anyone with an appropriate sounding title but someone at the corporation has the title "Assistant to the President," address yourself to him. It's a fairly safe bet he has something to do with company contributions.

In the case of federal government agencies, the *Catalog of Federal Domestic Assistance* provides the name and address of the appropriate person to contact under the agency listing. Always call first, however, to verify that the contact person is still connected with the federal agency in question. The same is true of state and local funders (many publish their own funding guides modeled on the *Catalog*), although personnel changes tend to occur with less frequency at the state and local as opposed to federal level. Your Congressman, community leader, or state librarian should be able to provide you with the name of an appropriate person to contact at a local funding agency, if that agency is not listed in any directory.

For small associations, clubs, societies, and other funders in the miscellaneous category, discussed in Chapter 5, simply write to the president or executive director as the name appears in the *Encyclopedia of Associations* or other directories or in the literature distributed by the organization itself. *Note:* Many funders in the miscellaneous category use printed application forms. So call or write for an application *before* you begin your proposal.

Title Page

The title page is optional and probably should not be included in a proposal of three pages or less. The title page helps make your proposal appear neat, well organized, and professional. It should include all of the following information in whatever order seems most reasonable to you:

- Amount requested
- Name of funding agency
- Purpose of grant (short descriptive title)
- Time frame of project (opening and closing dates)

- Your full name and affiliation, if any
- Your address and phone number
- Date of submission
- Name of your institutional or personal sponsor, if relevant

You may wish to experiment with various sequences and spacing of the above information until you arrive at an arrangement that is logical and visually appealing to you. The actual title that you select is important. It should be as short as possible while still relaying enough information to immediately give the reader a feeling for your project. Avoid abbreviations, acronyms, and "in" terminology in your title as well as in the body of your proposal.

ABSTRACT

In my opinion, every proposal should have an abstract or summary, even if the proposal itself is only a page or two long. In general, a short proposal entails a short abstract, but no abstract, however, should exceed 250 words. The abstract pleads the case for your idea in factual terms. It should be able to stand on its own as a clear, logical summation of your proposal.

Although the abstract is the very last thing you will write before putting the proposal package together, it may be the very first thing a funding executive will see. Depending upon the reviewing procedures of the particular funding agency, the initial impact of the abstract may determine whether or not your proposal lands in the reject pile because it is "outside the scope" of the funder. It is essential, therefore, that you submit an abstract as part of your proposal package. If you don't do this, you run the risk of having a staff member of the funding agency read your proposal and write an abstract himself. This could then be attached to your proposal and accompany it throughout the remainder of the reviewing process.

If precise writing is difficult for you, the abstract is one part of your proposal that someone else could help write. That other person, however, should be familiar with you and your ideas and should have read the proposal with great care. Remember to keep the abstract simple and uncluttered by excessive description. A good technique to follow is to write an abstract, then try to reduce it to one-half the number of words without distorting its meaning.

If your proposal is a long one, the abstract should be on a separate page immediately following the title page. If your proposal is short, the abstract may simply be part of your cover letter. You may wish to label it "Abstract" or "Summary," or you may include it as the first paragraph of your introduction.

APPENDIX

The key dictum about the appendix is to keep it relevant. Include only those explanatory or supplementary materials that help strengthen your proposal. If materials in your appendix seem truly essential to you, they probably should be inserted or at least referred to in the body of your proposal. If materials seem extraneous, they probably should be left out completely. Remember, appendixes add great bulk to your proposal. They make it seem lengthy and sometimes are not read carefully by grants decision makers. Always follow the specific instructions of the funding agency regarding the contents of your appendix.

Unless expressly prohibited by instructions of the funding agency from doing so, you should include in your appendix an up-to-date resume and a list of references. If your project is to be conducted under the auspices of a sponsoring institution, you also should include a copy of the tax-exempt certification (from the Internal Revenue Service) for that organization. This is important if the check is to be made out to a sponsoring organization or umbrella group rather than directly to you.

Following are some suggestions of items you may wish to include in the appendix to your proposal:

- Endorsement letters from qualified experts in your field
- News clippings or evidence of other media coverage of your work

- List of past support you have received from other funders
- List of awards or prizes you have won (may be part of your resume)
- Any state or federal certification that is required
- Brochures, reports, or other literature that you or your sponsoring organization have compiled describing your work
- Statistical data, including charts, graphs, and surveys, that help support your need statement
- Samples of your work if you are an artist, photographer, writer, etc. (Be careful what you submit. There is no guarantee that these samples will be returned to you.)

It is important to be selective about the contents of your appendix. Rather than including many items in your appendix, you might note that they will be made available to the funder upon request.

SUMMARY

To sum up, the entire proposal package consists of:

Cover letter
Title page* (optional)
Abstract
Proposal (see next section)
Appendix

COMPONENTS OF THE PROPOSAL

In most instances, the main body of your proposal should include the following components in the order in which they are listed here:

Introduction
Statement of need

---

*In a very short proposal (less than three pages), a title page may be unnecessary. If your proposal is eight pages or more, a table of contents should follow the title page.

THE PROPOSAL | 221

Objectives
Procedures
Evaluation
Budget
Future funding

These components answer the following questions:

Introduction—Who?
Need—Why?
Objectives—What?
Procedures—How and when?
Evaluation—How well?
Budget—How much?
Future funding—What will happen to you and your idea once the grant money runs out?

To further define the components of your proposal:

- The *introduction* helps establish your credibility as a grant applicant.
- The *statement of need* outlines a problem or points out a gap, explaining why you require a grant to solve the problem or fill the gap.
- The *objectives* refine your idea and tell exactly what you expect to accomplish in responding to the need.
- The *procedures* describe the methods you will adopt to accomplish your objectives within the time framework.
- The *evaluation* describes how you and the funding agency will measure the extent to which you have succeeded.
- The *budget* depicts in dollars and cents precisely how much money will be required and how and when it will be spent in order to accomplish your objectives.
- The *future funding* section predicts with some degree of detail the eventual outcome of the grant if awarded.

Your proposal should be written in such a way that there is a logical flow between the ideas, and one section leads to the next in a seemingly inevitable sequence. For each need you point out, there should be at least one corresponding objective with at least one matching procedure and one evaluation technique. You may find that you have more than one objective deriving from a single need, several procedures required to accomplish a single objective, and more than one test needed to evaluate whether or not you've succeeded. Or one procedure may accomplish several objectives, and so on.

You might diagram this part of your proposal along the following lines:

$$\text{Need} \longrightarrow \text{Objective} \longrightarrow \begin{array}{l} \text{Procedure I} \\ \text{Procedure II} \longrightarrow \text{Evaluation} \\ \text{Procedure III} \end{array}$$

or

$$\text{Need} \longrightarrow \begin{array}{l} \text{Objective I} \\ \text{Objective II} \end{array} \longrightarrow \text{Procedure} \longrightarrow \begin{array}{l} \text{Evaluation I} \\ \text{Evaluation II} \end{array}$$

You may find an example instructive at this point. Let us say that you are a sculptor applying to a corporation for a grant to create a sculpture whose function would also provide seating space in a local public park. The *introduction* to your proposal would state your credentials, institutional affiliations, if any, and why you are well qualified to create the sculpture. It would also mention that you have received permission from the appropriate municipal authority that oversees the park. The *need* section would indicate that the park, lacking such a sculpture, is underused because it is visually unattractive and has inadequate seating. Since the proposal is addressed to a corporate funder, mention specifically why corporate employees, consumers of corporate products or services, and members of minority or underprivileged groups would benefit from the creation of such a sculpture. This proposal would probably have two *objectives:* to improve the appearance of the park by installing the sculpture on or before a given date and to increase use of the park by installing the sculpture (including seating). The *procedures* section would provide a timetable for each stage, including purchase of tools and materials, the actual sculpting process, and installation of the finished product. The *budget* section would list in detail the costs of all materials and manpower (your own or perhaps that of an assistant if required for the installation). The *evaluation* section might include several tests: Did you erect the sculpture on schedule? Did you spend the amount of money requested? You might construct a survey of users of the park to determine their opinion of the appearance of the sculpture. In addition, you might perform an attendance check over a period of time to see if use of the park increases, especially among members of the population it is intended to serve. Finally, under *future funding,* you would note what maintenance costs, if any, would be required to preserve and maintain the sculpture. Perhaps the municipal agency that oversees the park would agree to take on subsequent costs.

As you can see from the above example, a relatively simple proposal, addressed to a specific funder, outlines one need, two objectives, several procedures, three or four evaluation vehicles, and the possibility of future funding.

The following is a more detailed discussion of the component parts of a proposal.

INTRODUCTION

The introduction to a proposal may be one sentence or a short paragraph or two. The purpose of the introduction is to establish your credibility as an applicant by indicating to the funder that you are a serious person of integrity, worthy of consideration for funding, and the logical one to carry out the project you propose. You may do this by providing a few relevant biographical details and by mentioning your sponsor, your mentor, or a well-known individual under whom you have studied or worked. A complete resume and list of references with full names, titles, and addresses should be added in an appendix to your proposal. Remember to follow the instructions of the funding agency in this regard.

Your introduction performs the function of suggesting to the funder that you are capable of success. After all, no one wants to back a loser. Consequently, previous successes should be mentioned. Contrary to what you may expect, referring to other grants or awards you have received helps rather than hinders your chances. It is a somewhat unfortunate truth that many funding agencies look for individuals who are safe bets rather than those who represent some risk. The fact that another agency or organization has previously funded you adds to your credibility as an applicant.

If possible, your introduction should briefly state why your idea is important, how it fits in with the grant-making goals of the funding agency, and why you are the logical person to receive the grant. The introduction should give evidence of your competence in your field and of the fact that you are fiscally responsible. If you have a sponsor, briefly describe your relationship to this institution and to your supervisor or mentor.

STATEMENT OF NEED

The statement of need tells why the funding agency should give you a grant. Usually, but not always, the need section of your proposal describes the service you will perform or the product you will provide and why the funder has need of it. This is your chance to portray the double value of your project: its intrinsic worth as an idea and its usefulness to society at large.

In your need statement, you may wish to indicate that you are aware of related work, research, or experimentation by other individuals in your field and the respect in which others presently are not satisfying the need you've identified. Showing how your own work fits in with others', while still emphasizing how it differs, suggests to funders that giving you a grant may have a beneficial impact on your field as a whole.

Hard evidence is required to back up your statement of need. It is not enough to identify a problem and to assume it is self-evident. You require documentation to show why it is a problem. Remember not to overstate your own personal need. Be sure to differentiate between your own financial requirements and the more acute need for a solution to a problem. Be careful not to describe a large, complex problem to which your idea offers only a partial, seemingly insignificant solution. Such overstatement can alienate the grants officer and consequently get you off to a negative start. Although documentation is important, use statistics sparsely and judiciously. Statistics do have their place, but they too can give the impression of overkill.

Make your need section as short as possible and keep it factual. Avoid emotional appeals. Leave out anything you cannot readily prove, and be prepared to prove each statement you make.

OBJECTIVES

The objectives section is truly the heart of your proposal. It tells what you intend (*not* what you hope) to accomplish. A good objective specifies a particular outcome within a given time frame at a minimum level of acceptability.[2] Thus, objectives are closely tied to specific procedures and to evaluation. Objectives should always be measurable. Measurable objectives imply by their very nature a built-in subsequent evaluation process.

Most proposals that are attractive to funders contain several objectives. Objectives should always be stated explicitly rather than merely implied. Objectives are often confused by proposal writers with goals or procedures. A goal is a long-term personal aim. A goal may not be measurable. It is more vague than an objective and usually cannot be accomplished within a given time frame. For example, getting together a book of your poetry is a goal. Having an anthology of twenty of your poems about nature published by Writers Publishing House by or before September 1983 is an objective.

Procedures, which are the methods adopted for accomplishing objectives, should never be substituted for objectives. Conducting an experiment is a procedure. Once you've done it, whether you have succeeded or failed, it is complete. On the other hand, conducting an experiment on May 29, 1984, that conclusively proves that a particular drug kills off certain bacteria is an objective. Many proposals are rejected either because they confuse procedures with objectives or because they set forth nebulous goals as objectives.

PROCEDURES

The section on procedures is likely to be the longest component of your proposal, since it describes in some detail the means you will use to accomplish your objectives. You may wish to state why you have selected these particular procedures as opposed to others. Indicate that you have considered alternatives but have chosen certain methods as the most likely to lead to success. Even those funding agencies that use standard application forms permit some leeway in the procedures section. Take this opportunity to show the funder that you are capable of ingenious, but rational, choices.

An important element of the procedures section is your description of a chain of events that follow one another in logical sequence. You should create a timetable with projected starting and completion dates for each phase of your project. Some may overlap, and the redundancy can be displayed by means of a chart or diagram. Take care not to underestimate the amount of time required to finish each part of your project. Evidence of realistic expectations on your part will have a favorable impact on funders. Sound planning is an essential aspect of your proposal.

The procedures section may include any or all of the following:

- Design—the unique or adaptable approach you've selected and why.
- Participant selection—who (other than yourself) will participate in the project, and why and how these participants were selected.
- Timing—how long you expect it will take to achieve your objectives and what the estimated starting and completion dates of each stage are .
- Administation—how your project will be run and recorded. How many and what type of reports you will make to the funder at what intervals.
- What type of financial records you will keep.
- Dissemination—how you will make the results of your project available to others so that society as a whole may benefit from your ideas.

The dissemination of the report on a successful project conceivably could be beyond the scope of the proposal and could hint at the desirability of future grants. Examples of vehicles for dissemination are reports, other publications, videotapes, media presentations, slide shows, press releases, training sessions, working models, and so forth.

The procedures component of your proposal allows the most room for you to exercise creativity. The methods you select are a reflection of your own ingenuity and may depict a unique approach to the problem at hand. Care should be taken not to permit your procedures section to become too detailed. The funder does not need to know how many pencils you will use per day nor when and by whom they will be sharpened. The funding agency is entitled, however, to a fairly accurate picture of how and when its money will be spent.

EVALUATION

To the inexperienced grant seeker the evaluation component of a proposal often presents a major stumbling block. Try to remember that evaluation is nothing more than a mechanical means by which both you and the funder determine whether or not you have accomplished what you set out to do. The evaluation component of your proposal simply suggests ways that you and your funder might measure the success of your project. This is an important part of your proposal, because it demonstrates that you realize the responsibility behind receiving a grant. Evaluation should be not just an afterthought. It should be built into the design of your procedures as a continuous monitoring system.

Keeping all this in mind, do not wrack your brains trying to come up with an entirely objective evaluation process. No evaluation is completely objective when human beings are involved. Some funding agencies use third-party evaluators;* others will evaluate you themselves. If

---

*For a listing of these, see the Foundation Center's *Directory of Evaluation Consultants* (New York: 1981). Most funders who use outside evaluators select them themselves. Occasionally, especially if your grant idea is highly technical in nature, you may be asked to recommend a third-party evaluator.

neither of these applies in your case, you should design your own evaluation component. The following are some examples of evaluation tools: observation by a disinterested person or specialist, interviews and/or questionnaires completed by those benefiting from your ideas, rating scales, standardized tests, checklists, attendance records, surveys, and longitudinal studies. The list could be continued according to your needs and ingenuity. From the point of view of the funding agency, it is your own good faith and the funder's appraisal of your probable success that are at issue rather than which evaluation procedure you select.

The section of your proposal on evaluation attempts to tell how well you succeeded in accomplishing your objectives. It also answers the following questions:

- Did you operate as intended, following the methods outlined in your procedures section?
- What beneficial changes have been brought about that are directly attributable to your idea?
- Is your idea the only variable responsible for these changes?
- What conclusions may be drawn from this evaluation?
- What future directions may be projected for your field as a result of your accomplishments under this grant?

BUDGET

How much do you need? As mentioned earlier, determining precisely how much to ask for is difficult for the novice grant seeker. The amount you would *like* to get and the size of the grants a funder typically gives will naturally affect how much you will ask for.

In the area of government funding, in particular, there has been a regrettable tendency among some grant seekers toward applying for "overblown projects."[3] This has been due to a prevailing misapprehension that large projects tailored to the funding agency's own interests are more apt to get funded than smaller ones, closer to the individual's own heart. As a consequence, the grant recipient may find himself conducting a project only marginally connected with what he originally set out to do. Also, he will have to spend considerable time administering a large grant project whether he be an academic, an artist, a writer, or a researcher, untrained in the skills of a financial supervisor.

To avoid this trap, be honest with yourself. Ask yourself what's the *minimum* amount of support you need to start and to work on your grant idea for one year. Most grant projects are of one year's duration; for some it is possible to reapply before the year is over. Certain projects might be accomplished within an even shorter period of time. Others require one-time expenditures for supplies, equipment, space, or materials. For these kinds of projects, it is relatively easy to come up with a figure in dollars.

If you're still having trouble determining how much grant money to ask for, check out your field. What is the scope of similar projects conducted by your colleagues? What is the size of the average grant awarded by funders in your locale or subject field?

You might, of course, feel tempted to fabricate an attractive figure. This kind of ploy, however, may very well backfire. If you do not have a specific rationale for the amount of money you're requesting based either on the requirements of your particular idea or on the size of the grants funders ordinarily award in your field (if the two coincide, even better), your chances of actually getting a grant for that amount are practically nil. Grants are based on need; they are rooted in reality. Hard data and cold facts are required, not idle wishes.

Once you've arrived at a reasonable figure, you are ready to proceed with the actual construction of a budget. *Never* submit a proposal without a budget unless the funding agency expressly instructs you to do so. Some grants decision makers turn first to the budget before even looking at the rest of your proposal. The very first question raised by these execucutives will be: Is it reasonable? Don't ask for more or less than you really need. Padding is immediately discernible to the experienced grants officer, and too lean a budget suggests poor planning on your part.

Your budget is merely a picture in dollars and cents of the ideas set forth in the rest of your proposal.[4] There should be a relationship evident between the expenditures allotted in the procedures section and the budget, which ordinarily follows either the procedures or evaluation sections. Depending on the type of evaluation you recommend there may even be a budget item to cover its cost. By far, the most important connection is

between the objectives of your proposal and the budget. The grants officer will want to know if the total dollar amount you require is commen-
surate with expected results. In other words, is it worth the effort?

If you already conduct your day-to-day financial affairs according
to a budget, designing one for your grant idea will be easier for you. If
not, this is a good time to get started. For help in this regard, you may
wish to see Stephanie Winston's book, *Getting Organized,*[5] in which she
recommends a financial master plan for individuals wishing to follow a
budget. The plan includes fixed and flexible expenses and is adjustable
year to year as your financial picture changes.

If the grant for which you are applying includes living expenses, you
will have to spend some time determining what it truly costs you to live
in the location where you will work. A complete record of methods used
to compute each budget item should be kept on file to show how you
arrived at each figure. In case you are asked, be prepared to support each
budget item in minute detail, although the actual budget you submit
need not include all of this support data. On the other hand, expensive
items, especially travel, should be fully annotated within the text of your
budget.

If your project involves others who are either volunteering their services or salaried, you will need an entirely different type of budget. For
help, you should refer to texts on budgeting for nonprofit organizations
or ask an accountant or bookkeeper friend for assistance. The typical
budget *for an organization* ordinarily is divided into two sections: person-
nel- and non-personnel-related expenses. Your budget might include the
following items:

Personnel-related	Non-personnel-related
Salaries (your own and/or that of your assistants, usually prorated)	Space
	Equipment
Fringe benefits (if any)	Consumables
Consultants and contract services	Travel
	Telephone
	Other (postage, printing, fees, dues, etc.)

For the small grant project, it is conceivable that you will have little or nothing in the "personnel" category.

Try to ascertain from the funding agency how detailed a budget is required and get competent help in writing a budget whenever possible. One of the best resources in this regard is a friend or acquaintance who is already a successful grant recipient.

FUTURE FUNDING

Unless your idea requires a one-time expense and can be fully accomplished within a given time period, your proposal *must* include a section on future funding. As mentioned earlier, no funding agency wants a recipient who will become a parasite. Funders want to know how your idea will be funded, if necessary, when the grant for which you are presently applying elapses. It may very well be that you will be returning to the funding agency for a renewal. If this is likely, be honest about it, but suggest other possibilities as well. Read the guidelines of the funder to ascertain in advance what their policy is on renewal grants.

Keep in mind that future funding is a plan, not just a promise.[6] This section of your proposal may be short, but it should be specific. If possible, avoid vague allusions to future developments. Before you sit down to write this section of your proposal, draw up a self-support plan, including all the ways your idea might adequately be funded once you've used up the grant money. Then select the most likely for inclusion in this section of your proposal. (See Chapter 8 for a discussion of alternative sources of funding.)

Your future funding section should be detailed and well thought out. Above all, it should seem feasible to the funding agency. Remember funders like to feel that they'll be able to get in and out of your project quickly and with ease. They don't want to leave you dangling at the end of the grant, nor do they want you dependent upon them forever.

STYLE

Very little has been written on the suggested style of a proposal. Virtually no funding agencies provide useful information on this in their guidelines. It may be argued that style is not all that important as long as the proposal contains the correct information that the funder requires. I believe, nevertheless, that the style in which your proposal is written can have a decisive impact on whether or not you receive a grant, especially in circumstances where grants officers are deciding between two equally qualified applicants. Hence, style is worthy of discussion here.

Always begin with an outline. Your proposal probably will go through several drafts before it is completed. With each draft, pare down your sentences, eliminating excess verbiage. Clarity and conciseness are essential, so don't use two words when one will do. Shun technical jargon or "in" terminology; it may be "out" by the time your proposal is read. Assume that the funding executive who will read your proposal is a lay person without any particular expertise in your field. Steer clear of abbreviations and acronyms whenever possible. If you must use them, explain them fully.

It may seem self-evident, but good grammar is important. Don't use sentence fragments, and try not to let your thoughts run on unnecessarily. Employ correct spelling and proper punctuation. Your proposal should be composed of complete sentences, each expressing a distinct thought. The text should be divided into short paragraphs with headings and subheadings employed for clarification. If you feel insecure in this area (many artists and inventors as well as other nonacademics do), ask a friend, relative, or colleague with a strong writing background to scan your proposal for errors in grammar or syntax before you submit it to the funder.

The tone of your proposal is also important but difficult to define. A good suggestion to help you strike a suitable tone is to reread any literature, annual reports, or program guidelines provided by your funder immediately before you sit down to write. You may find that subconsciously your tone will tend to be consistent with the tone of the materials that you have just read. Try not to be either too stuffy, too casual, or too self-conscious. Address yourself to a human being, not to an abstract entity. Without resorting to emotional arguments or subservience, try to involve the funder in your project. The tone of your proposal can help establish that all-important link between you and the funder.

LENGTH

I am frequently asked how long a proposal should be. The answer I usually give is "long enough." By this I mean to say that your proposal should be as short as possible without leaving out valuable information required by the funder. There is room for wide variation here depending on the amount you're requesting, the nature of your project and the funding agency to which you are applying. Always ask the funder what length is recommended if this information is not provided in the printed guidelines.

Speaking in a general way, I find that one page is usually too short and more than five pages usually too long for all but the most complicated individual grant projects. If your proposal tends to be lengthy, the abstract is essential. As already mentioned, abstracts should be not more than 250 words.

SHORTCOMINGS OF MANY PROPOSALS

Although I dislike stressing the negative, you should be aware that a large proportion of proposals are turned down. Some say as many as 90 out of every 100 applicants are rejected (depending upon the nature of the grant),[7] though it would be very difficult to verify this statistic. You should not be discouraged by the high rejection rate. If you were to examine the rejected proposals carefully, it probably would become evident that most of the applicants simply had not done their homework properly. All too often their proposals were not worthy, inadequately prepared, or thoughtlessly submitted to inappropriate funders.

My research indicates that, like successful proposals, rejected proposals have much in common. Funding executives report the same reasons for rejection time and time again. Note should be taken of the following shortcomings so that you can avoid these pitfalls yourself.

TEN REASONS PROPOSALS ARE TURNED DOWN

1. Inadequately presented statement of need—perceived by funder as either not a significant issue or as one of such magnitude that a few grant dollars would barely make a dent in the problem.
2. Objectives are ill defined—put forward as vague goals or as personal aims.
3. Procedures are confused with objectives.
4. Lack of integration within the text among components of the proposal.
5. The funder does not accept proposals from unaffiliated individuals.
6. The funder knows that the proposed idea has already been tried and failed.
7. The funder approves of the concept but believes that the applicant is not the proper individual to conduct the project or that the institution with which the applicant is affiliated is not suitable.
8. The individual has adopted a poor approach and appealed on an emotional or a political rather than a factual basis.
9. The idea costs too much.
10. The funder does not have enough information.

Shortcomings 1 through 4 will be avoided as a matter of course if the simple recommendations provided in this chapter are followed. To prevent rejection due to shortcoming 5 see Chapter 2 on how to affiliate with a sponsor. Shortcomings 6 and 7 are more difficult to circumvent. However, if you survey your field carefully in advance and develop a thorough familiarity with the activities of others in your area, you may be able to prevent rejection on these grounds. If someone has already tried your idea and failed, it might be well to abandon the whole project. If you elect not to do this, then you should indicate to the funder why you have a better chance of success. Shortcomings 8 and 10 are inexcusable and demonstrate carelessness and a lack of common sense on the part of the applicant. Shortcoming 9 is the most difficult to avoid. The opinion that something costs "too much" is truly subjective and leads you to ask: relative to what? Perhaps the best caveat in order to preclude this reason for rejection is to present as tight a budget as possible, free of padding, yet adequate to permit the attaining of the desired results.

IMPROVING YOUR CHANCES

No one can guarantee that your idea will be funded. There are, nonetheless, a number of common sense rules that, if followed, should increase the odds in your favor.

Examine every publication available relating to a potential funder with minute attention to detail. Follow up every lead. Before you sit down to write a proposal or to fill out an application form, ask yourself one more time: Do I have the most up-to-date information available on this funding agency? Have I done everything to ascertain what format and procedures they prefer?

Never bypass important executives or program officers by attempting to use your influence. This is particularly true when applying to foundations. Knowing a board member or having political connections certainly cannot hurt your chances, especially if the funding agency has no full-time staff. It is essential, nevertheless, that you go through channels and play the grant-seeking game by its rules. If you have some influence with a particular funding agency, you should exploit that advantage subtly and with discretion.

In all correspondence and meetings with potential funders, be as noncontroversial as possible without compromising your ideals. This does not mean that you should misrepresent yourself in any way. It is simply intended as a reminder that to a funding executive you are an investment for which he may be held accountable someday. In most instances, it is easier for the funder to justify giving you a grant if you appear solid, stable, and generally acceptable by society's standards. Yet, there are glaring examples when conformity is not a virtue. Be sure of your grounds in such instances.

Keep all promises to a funding agency; don't make those you can't keep. It is important that an atmosphere of mutual trust be established between you and the funder. Reliable, predictable behavior on your part during the initial stages of application helps to reinforce the impression that as a potential grantee you are a reasonably safe risk. A display of realistic expectations about what you can achieve bolsters the funder's confidence in you.

Cover your bets. Submit all information required in exactly the number and format requested. Respond promptly to all phone calls and letters from the funding agency or its representative. Rehearse all presentations. Decide in advance what you'd like to convey during an interview, if you're able to procure one. In other words, be prepared!

Comply with all laws, rules, and regulations established by a funding agency. Follow application guidelines to the letter, even if they appear nonsensical to you. Do not submit more or less information than is required without a very good reason.

Finally, never use someone else's proposal. This point cannot be overemphasized. When a funding agency gives you a grant, the decision makers involved are supporting both your project and you as a person. Someone else's ideas will not help to present you in a better light. Don't forget, the world of grant making is a surprisingly close-knit one. If you usurp someone else's proposal, the chances are that you'll get caught. Although this is not necessarily illegal, it can result in a bad reputation for you among important funders. Don't take the chance!

SUBMISSION TECHNIQUES

No discussion of proposal writing would be complete without some advice about what to do with your proposal once it is written. The assumption is made, of course, that you already have selected an appropriate potential funder, determined when the deadline is, and whether a personal interview is possible prior to submission. Also, you will have secured copies of all printed leaflets, reports, and guidelines distributed by the funder. (See Chapter 6.)

Checklists are useful to ensure that all items required by the funding agency are included in the proposal package and in the proper order. Be sure you have two or three extra legible copies of the proposal in addition to the number of copies required by the funder. Double check one more time to be sure that your cover letter is addressed to the correct person with appropriate title at the right address. If your are applying for a federal government grant, you may have to submit copies of the proposal to a local, regional, or national office. Find out which one.

The very last thing you should do, immediately prior to submission, is to have someone else, a colleague, co-worker, or acquaintance who may have received grants himself (a friendly funding officer would be nice, if you know any), read and review your proposal. Any author can tell you that sometimes you get so close to what you have written that you are unable to truly read it anymore. When this happens, even glaring errors or spelling mistakes may be undetected. Someone in an unrelated field might catch any jargon you have fallen into using. So it is a good idea to have someone whose opinion you respect read the proposal one last time before you send it off.

This brings us to the next issue—paragraphs and paragraphs in fund-raising guides have been devoted to the questions—should you mail your proposal, hand deliver it, or send it by messenger. Should you try to find someone "important" or influential to transmit it for you? Although the issue of just how to get your proposal into the funder's hands may seem to have peripheral importance to the grant seeker, it is hard for me to believe that the manner in which a proposal is submitted can have much of an impact on the ultimate decision whether or not to fund it. Rather, you should concern yourself with how to ascertain whether or not the funder actually has received your proposal and what action (if any) is being taken on it.

If the funding office is located in your own or a nearby city, you might be tempted to deliver the proposal in person. After all, this is the only way to truly guarantee that it reaches its destination. Also, you might harbor the hope that just as you arrive at the funding office, an important grants decision maker will be on his way out to lunch. Having met you, he might be more receptive to your ideas. In any case, delivering your proposal yourself certainly can't hurt. It gives you a chance to view the funding office and to meet the receptionist or secretary.

If your concern is ensuring that your proposal arrives intact, you might decide to send it by messenger, registered mail, or "return receipt requested." Sending it this way may satisfy your anxiety. However, along with delivering it yourself, this form of transmittal has a major disadvantage. It does away with your one good excuse for contacting the funder several weeks hence—to find out if, in fact, your proposal arrived intact!

The Waiting Game

One of the most difficult jobs for the novice grant seeker is waiting to hear the funder's decision. Yet once the proposal has left your hands, there is little you can do but wait. You may have expected to feel a great sense of relief once the proposal is submitted. Instead you find yourself terribly anxious. After several weeks have elapsed and you still haven't heard anything, your fingers may begin itching for the telephone dial. The temptation is great to place one quick call just to see if they got the proposal and how things are going.

If you must, go ahead and call. Ask what you have to ask and then hang up. Quit while you're ahead. Don't go on and on about your grant idea on the phone. On the other hand, be prepared to discuss your proposal, on the remote chance that you do get a grants officer on the phone. In all circumstances, avoid any behavior that may be construed as nagging.

I personally believe it is a bad idea to keep calling and checking with the funder about the status of your proposal. Your initial planning should have taken into account a substantial block of time for this waiting period. Even funders of small awards and prizes may take several weeks to decide. Corporations and foundations may require several months to complete the decision-making process. Some government agencies take even longer. (Five to nine months is the norm.[8]) After an acceptable time has elapsed (let's say one month to six weeks), you might wish to make a discreet call or send a polite note inquiring about the status of your proposal, whether funding officers need further information, and when you will hear the results. However, it is far better to ask at the time of submission about how long the process will take and when you might expect to hear their decision. Once this time period has elapsed, *then* it would seem to be acceptable to call again.

Another temptation the grant seeker faces is using his connections (if he is lucky enough to have them) to apply subtle pressure on the funder or even to submit the proposal for him. Those who believe in submitting preliminary letters of inquiry often recommend the use of influential Congressmen, friends of corporate board members, or foundation trustees to make the initial inquiries or to check on the status of the request. Earlier in this chapter, I mentioned that with most funders, it is best to go through channels. If you know someone who can help get you a grant, now is the time to make use of such a connection. However, do so with great discretion. Tread lightly so as not to injure anyone's feelings. Receptivity to influence varies widely depending upon the funder's own giving styles. The moral is—know your funder.

APPLYING SIMULTANEOUSLY TO MORE THAN ONE FUNDER

The discussion of submission techniques presupposes that you have a potential funding source to send your proposal to. As mentioned earlier, most proposals are written, at least in part, with specific funders in mind. To answer one last question that troubles many grant seekers, it is perfectly all right to submit the same or a similar proposal to several funders simultaneously just as long as you are totally honest with each funder that you are doing so.

A good suggestion is to mention in your cover letter that you are also applying to *a, b,* and *c* funders. The grants officers in question probably will find out about your other applications anyway, when your references are checked. Stating this up front helps to reinforce your personal integrity and really can't hurt your chances.

Your concern might be that in acknowledging your simultaneous application to several funders, they may use this as an excuse to lay your proposal off on one another. However, if the grant idea is attractive to funders, the reverse also might happen: it might stir up their competitive instincts and cause one of them to act more quickly on your application.

What if two funders both award you a grant? Don't worry about this unlikely possibility. With grant seeking as competitive as it is today, I doubt this will happen. If it does, you'll just have to decide which one to accept. Maybe you can find a way to enlarge upon or extend your idea to incorporate additional grant money. Perhaps both funders can combine their resources to support you jointly. If not, then warmly thank the funder whose grant you can't use right now, but leave the door open for future support. You probably will be back to the second funder at a future date.

DEADLINES

Deadlines are mentioned in various contexts throughout this book. It seems almost superfluous to repeat this one more time but—deadlines are important! So observe them. Don't send your proposal in too late. This is a surefire way to jeopardize your chances.

If you can't ascertain what the deadline is, then submit your proposal as early as possible. I have head of instances where early submission increases your chances. Some funders operate on a more or less continuous granting schedule at certain times of the year. Once their funds get low, they become more selective. Also, don't be a "deadline squeezer."[9] Send in your proposal at least two weeks in advance of the deadline, earlier if possible. This enables the funder to request additional information if needed, and it gives you a chance to put the necessary information together.

NOTES

[1] Conrad, D. L. *The grants planner.* San Francisco, Calif.: The Institute for Fund Raising, 1976.

[2] Hall, M. *Developing skills in proposal writing* (2nd ed.). Portland, Oreg.: Continuing Education Publications, 1977.

[3]Whyte, W. H., Jr. *The organization man.* New York: Simon & Schuster, 1956. P. 236.

[4]Dermer, J. *How to write successful foundation presentations.* New York: Public Service Materials Center, 1975. P. 80.

[5]Winston, S. *Getting organized.* New York: Norton, 1978. Pp. 99–101.

[6]Kiritz, N. J. Program planning and proposal writing. *The Grantsmanship Center News,* January 1974, P. 14.

[7]Lee, L. *The grants game.* San Francisco: Harbor Publishing, 1981. P. 55.

[8]Hillman, H. *The art of winning government grants.* New York: Vanguard Press, 1977.

[9]Glass, S. A. A winning strategy for victory in the home stretch: Spend more time cultivating foundations and less mailing proposals. *Case Currents,* May 1980, p. 30.

"If its GO and you succeed—smile a lot. You've earned it. And get to work on your next idea. If it's NO GO and your idea fails—cry a little. You've earned it. And get to work on your next idea. Good luck! From those of us who have succeeded. From those of us who have failed. From all of the above."
—DON KRACKE WITH ROGER HONKANEN
How to Turn Your Idea into a Million Dollars

Chapter 8 / FOLLOW-UP

IF YOU RECEIVE A GRANT . . .

Let's assume that your grant-seeking endeavor has had its hoped for result: you've received the grant you applied for. At long last your idea is funded! You may breathe a sigh of relief and pat yourself on the back. However, this is far from the end of your relationship with the funder. In many ways, it is just the beginning.

Throughout this book I have touched upon the concept of an alliance between the grant recipient and the funder. Money is rarely given away and forgotten. Instead, a kind of mutual agreement almost automatically takes shape between you and your funder. This alliance suggests some form of repayment on your part. Precisely what is expected in return for the grant may not always be clear at the outset. (You should endeavor to make it so.) However, hardly ever is a grant given without the implicit concept of a loan or investment.

Let us stand back a minute and consider the actual steps involved in receiving and managing your grant dollars.

How Will You Be Notified?

If you do receive a grant, you, in most cases, will receive a formal letter stating that you are being funded, accompanied by or followed under separate cover by the check itself (most likely for the full amount). In some rare instances, you simply may receive a check in the mail. More often, there will be some time lag (from several weeks to several months) between official notification of funding and the actual receipt of a check. This lag is characteristic of foundations and some state and local agencies. Corporations and federal government agencies tend to be more prompt about payment. Occasionally, you may hear first by telephone, or if your grant is from a federal agency, your Senator or Congressman may be notified directly, and word may come to you from his office.

There are three points to keep in mind about the notification that you are funded:

1. Never accept a verbal okay to go ahead with a grant project. Always get it in writing. Politely but firmly request a written confirmation on formal letterhead stationery. Never spend any of your own money until you have the grant check in hand.
2. Promptly write and thank the funder for awarding you the grant. Address yourself to the head of the funding organization. However, mention by name decision makers who spoke with you and/ or interviewed you and who may have been instrumental in getting you the grant.
3. Be sure you understand all conditions tied in with the grant notification. Are you aware of everything that will be expected of you in terms of carrying out your grant idea?

Now is the time to make expectations explicit between you and the funder so that no one will be disappointed at the completion of your grant project. The very first thing you should do upon official notification that you are funded is to find out about any and all strings attached.

QUESTIONS YOU SHOULD ASK

Before you begin to expend the grant money, you should resolve the following questions:

Is the support available sufficient for your needs? Whereas in another chapter I criticized the tendency to apply for more money than you actually need, it can also hurt you to accept a grant that is not large enough. Accomplishing only part of your grant idea, or doing it in a way that cuts too many corners may be detrimental to you, your idea, and your reputation. In some instances, if is better to refuse the partial funding and to keep looking. Some grant seekers subscribe to the school of thought that favors beginning with whatever they get, hoping to be able to wangle more later on. However, I believe it's almost always better to go for the full amount, if possible.

What are the actual terms and conditions of the grant? Are they conveyed in a written letter of agreement? Are there requirements attached that you are not prepared to meet? Remember that a promise, be it verbal or written, may be considered as binding as a contract by some funders.[1] Don't make any promises you can't keep. Accepting a grant should never force you to compromise your own ideals or behave in a manner that is unethical, illegal, or contrary to your own convictions.

Are there restrictions on the use of the money? What types of expenditures are not allowed? Can the grant dollars, for example, be used only for direct costs of running the project and not for salaries or other types of expenses? This particular restriction is fairly common with foundation grants and not always made explicit by the funder up front. Be sure to find out about it in advance. Also, determine how much latitude you will have in making your own administrative decisions without going back for an okay from the funder.

How will the grant be paid? In a single check or by multiple payments? Will the check be made out to you directly or will you receive the funds via your sponsor?

What kinds of records (financial and otherwise) will you have to keep? What receipts will you have to save? What vouchers and other forms will you have to fill out as your grant project proceeds?

What kinds of reports will you have to make to the funder? How long will they have to be? Some funders have fairly strict requirements about reporting. Be sure you understand them.

How will your work be evaluated, by whom, and how often? Should you expect an audit? As mentioned in the chapter on your proposal, there may be two types of evaluation of your grant project: the one you conduct yourself and the one conducted by your funder. It is important that you clarify evaluation procedures in advance so that you can plan accordingly.

Are renewals and/or extensions possible? If so what are the procedures? Some individuals begin applying for an extension or renewal almost immediately upon the receipt of the initial grant money. Although I don't recommend this practice, in some cases where funders only make awards one year at a time, and your idea requires several years' funding, it may be necessary. The point is to be sure that you can complete the project with the funds allotted *and* within the time allowed or that applying for more money and/or time will not present a problem at a later date.

In some instances (e.g., grants from certain government agencies), if the money is not spent by a particular date, the balance must be returned to the funder. To avoid this eventuality, find out about expenditure requirements at the very outset of your grant project.

GRANT ADMINISTRATION—YOUR RESPONSIBILITY

There have been numerous books written on grant administration, albeit mostly directed to the leaders of nonprofit organizations. It is not the purpose of this chapter to treat this subject in a detailed way except to suggest some basic principles and to remind you that accepting a grant is a serious business. No matter how small the amount in dollars or who the source of the original funding, receiving a grant represents a responsibility on your part that is not to be taken lightly.

If you are affiliated with a sponsor, you immediately should make an appointment with the fiscal officer of that institution to determine what accounting procedures must be followed. Financial record keeping for your grant project should be in accord with that of your sponsor while still satisfying all the requirements of your funder. This may be trickier than you think, especially if your sponsor's and your funder's fiscal years do not coincide.

As mentioned earlier, if the grant is small, the tendency among individual grant recipients may be not to take the matter of financial management seriously. Avoid this mistake. Even the smallest of federal government grants may be subject to an audit; private funders may demand repayment if it is believed that you have misused the grant funds. To protect yourself against such possibilities, however remote their occurrence, conduct your grant project as if at any moment you expect an audit.

Keep complete and permanent records for your own taxes. Remember, many awards as well as payment for work under private grants and government contracts are treated as taxable income by the IRS.*

Prepare and file your reports on time. Keep continuous records in a businesslike fashion. Some funders share the stereotyped opinion that the more creative the individual, the less reliable he will be in handling and reporting his finances. You can help combat this myth by your own exemplary behavior as a grant recipient. It will also help you when you ask for future grants.

For the duration of the grant project, keep everything of use together, such as vouchers, statements of services, travel, or other receipts. Stapling related materials will make your life easier later. Keep your own daily or weekly diary of grant-related matters, plus monthly or quarterly records for the funder, depending upon reporting requirements. Write legibly and neatly, even though no one else but you probably will ever read these notes. There is nothing more frustrating than your own notes that prove incomplete or illegible when the time comes to refer to them.

As already noted in the chapter on getting into gear for grant seeking, the fewer the files you keep the better, and alphabetical filing arrangements seem to work best.

---

*To find out more about this, see *Scholarships, Grants, Stipends and Taxes, Too* (Chicago: Commerce Clearing House, Inc., 1977, or latest edition) and if you are an artist, *Fear of Filing* (New York: Volunteer Lawyers for the Arts, 1981, or latest edition).

By far the best advice I can offer you on grant administration is to act in a professional manner. To quote one corporate grant recipient, "excellence and success create a substantial base for the future."[2] Your behavior as a grant recipient may have a real impact not only on your own receipt of future grants but on the fate of other grant applicants as well.

RELATIONS WITH FUNDERS

I mentioned at the outset of this chapter that once you receive a grant, your relationship with the funder is really just beginning. This special relationship serves as a form of reinforcement to assure decision makers that they did the right thing in awarding you the grant, and it also enables you to ask for help or advice as you need it.

You should keep up a continuous communication with your funding representative throughout the year. If it is not obvious or stated in your grant notification who your representative is, you should find out who will serve as your grant contact or liaison. Introduce yourself by phone or in person to this individual and make your intention known of keeping in touch with him *on a selective basis* throughout the duration of your grant project.

Aside from the formal reports you are required to submit, you may want to keep up more informal contacts by correspondence or telephone with your grant liaison, informing him on a periodic basis of your activities, accomplishments, unexpected difficulties, and solutions, as well as other grants, awards, recognition, and praise you may have received along the way. Make him feel like an integral part of your grant project. He indeed is: without him your idea might well still be in the planning stages. So try to keep your relations with the funding contact friendly.

In general, it is best to let the funder take the lead in initiating most communications with you. However, one or two unsolicited brief progress reports or letters of support for what you are doing during the course of the year, submitted to the funder just to show how his money is being spent, is not a bad idea.

Put funding executives on your mailing list; simply remember to be selective about what you send them. If your project involves an opening, a performance, or demonstration, you definitely should send funding executives invitations to attend. Other on-site visits should be initiated by funding executives themselves. Remember that thoughtfulness is just good business in grantsmanship as in any other field.

A word of caution, however, avoid inundating funding executives with press releases, other written materials, phone calls, or requests for advice about how to proceed on trivial matters. Too much communication with funders can have as adverse an effect as too little. Your relationship with your funder should be concentrated on achieving four objectives:

1. Showing your appreciation
2. Cementing the alliance between you
3. Proving he made the right judgment in selecting you and your idea
4. Paving the way for extensions, renewals, or future funding, if needed

Accomplishing these objectives are the actual challenges in grant administration.

Dissemination

There is one element of grant administration that sometimes presents a thorny problem to grant recipients—that is the dissemination of information both on the receipt of the grant money and on the outcome of your grant project. On the one hand, you may view it as your responsibility to inform your colleagues, other grant seekers, and perhaps even the general public about the development of your idea. A theme throughout this book has been the sharing of information and resources, in order that one individual may build upon another's successes.

On the other hand, there has been a long-standing proclivity for avoiding the limelight among funders, especially private foundations and corporations. The hesitation of funding executives in regard to dissemination of information on their grant awards is somewhat understandable. Funding executives fear a barrage of requests they can't fill, too much scrutiny by the IRS or other watchdog agencies, and public criticism or scorn if the grant projects they fund are less than completely successful.

The desire to keep grant awards secret is an outdated attitude, and pretty much impossible, anyway, in light of the recent establishment of computerized files, periodicals, directories, and other resource materials that regularly report to the public on foundation, government, and to some extent now even on corporate grants. Nevertheless, tread lightly in the area of dissemination. Find out at the outset of your grant project about your funder's policies on dissemination of information on your grant and about publication of findings, if any.

Clear all press releases, fliers, promotional publicity, and so on with your grant liaison at the funding agency before disseminating information to the media or to the general public.* Although you might see this as censorship, it is really just good business sense and indicative of responsible behavior on the part of the grant recipient.

IF YOU ARE TURNED DOWN ...

FIND OUT WHY

If your proposal is rejected, the very first thing you should do is find out why. Be sure you thoroughly understand the reasons. You may write a letter thanking grants decision makers for their time and requesting a written explanation of why you were turned down. A more direct approach, however, is to call and ask for ten minutes of the funding executive's time to find out why. Don't be shy about this. Explain that you are *not* trying to change their minds. Rather, you simply value constructive criticism to guide you in reshaping your proposal and in resubmitting it to funders in the future.

Some funding officers will discuss why they turned you down over the phone (a few will even do so in person) as long as you make it clear

---

*The reasons to send out press releases, fliers, or other promotional literature about your grant project would be to achieve public recognition, enhance your own career, build up a clientele, or attract other new funders.

that you are seeking information, not reconsideration. Staff members of some government agencies will provide you with a "debriefing session" regarding what special factors caused you to lose the grant, award, or contract. The provisions of the Freedom of Information Act obligate the federal government to reveal the reasons why any grant proposal was not funded.

LEARN FROM YOUR MISTAKES

Persist until you're sure you understand why you were rejected. This information is valuable to you since you want to learn from your mistakes—the cardinal rule in responding to rejection. Find out if you can resubmit a revised proposal or reapply with the same or a new proposal during the next grant-making cycle.

Remember, the reasons a funder says no today may have little to do with you or your idea. The funder's own budget may be cut or all grant money expended for the current fiscal year. There may be internal management problems at the funding agency. Someone else from your geographic area or field of endeavor currently may have a grant from this particular funder. They may be concentrating right now on certain limited aspects of grant making.

None of these reasons preclude your reapplying six months or a year from now. So be sure to ask about this possibility. The main principle about handling rejection is to ask yourself: What have I learned that I can use next time?

REFERRALS, REFERRALS, REFERRALS

Just as soon as you learn why you were rejected (as long as the reason has nothing to do with the funder finding fault with you or your idea), ask immediately of the grants decision maker who else might fund you. Remember, you have an important advantage at this point in time: Funding executives may be feeling a bit guilty about you and consequently more willing than usual to refer you to their counterparts at other funding institutions.

After learning from your mistakes, the second principle in responding to rejection is always to obtain a referral! Some funding executives may even be willing to introduce you to others if they like you and/or your idea. In any event, it can't hurt to ask. Proceed on the premise that if they can't give you the money, at least they can tell you how to get it from someone else. But do so politely. You want to gain a reputation among funders as an assertive, sincere, and responsible grant applicant, but beware of the labels "pushy" or "obnoxious." They tend to stick once they are attached to you.

You Can Get More from Funders Than Just Money

This is the third principle of responding to rejection. It is a serious mistake for the grant seeker to view funding executives merely as dispensers of dollars. It is a very narrow view and self-limiting besides. As we've already noted, funding executives who've just turned you down can provide valuable referrals and introductions to others. Furthermore, they may contribute sound ideas about how to improve your proposal, implement your grant idea, and fund it in the future.

Those funders who've rejected you and those who give you grants may provide a whole range of additional services as well. Many of these services fall under the heading of "technical assistance." They may include actual assistance in writing or rewriting your proposal, identifying other funders if this one is inappropriate, providing printed guidelines, other materials, and hints on the review process, help with money management and administrative decision making, public relations for improved credibility, evaluations, on-site visits, and access to contacts for joint funding.

So leave the lines of communication open between you and funding executives. If you handle the rejection properly, you may make a positive mark in the funding executive's memory that will prove valuable in the future. Treat all funding executives with respect, and try to capitalize on this experience for future gain.

ABANDONING YOUR IDEA

There is another possibility that cannot be overlooked. Some grant ideas are unlikely to be funded because the timing isn't right, the geographic location is inappropriate, money simply is not available, or due to other external factors beyond your control. For example, your idea may be good, but someone else already may be doing it. This brings us to the fourth and last principle of responding to rejection of your grant idea: know when to give up and proceed on to a new idea or project.

There is no shame in giving up on your grant idea once it truly appears that it will be impossible (or even extremely difficult) to fund. As we discussed in an earlier chapter, certain ideas turn out to be unattractive to funders at a given time, even though they may seem valuable to you. Deciding at some point not to go ahead and seek a grant is a legitimate choice and a viable option for the grant seeker. There are other options as well: you may try to fund your idea another way (see the next section) or temporarily shelve it for later on when conditions are better and you have more time to devote to it.

Rejection is all in your point of view. It is possible to see the difficulties you encounter in getting funded as challenges to your own creativity and ingenuity. Try not to take criticism too personally. Once you expose your grant idea to funders, it is irrevocably set out for public judgment. Of course, you'll always feel some anger over the criticism by others of something so close to you. However, learn to separate your own identity from the essence of your idea. Rarely will funders be turning you down for personal reasons. Rejection almost always has more to do with the nature of your proposal or with other grant-making factors. When you learn to accept this principle, living with rejection by funders will become that much easier.

ALTERNATIVES—OR WHAT TO DO WHEN THE GRANT MONEY RUNS OUT

Once you've come to terms with the rejection of your proposal or, in some instances, once the grant money runs out, and your idea is still not fully accomplished, you are left with the decision of what to do next. You can always reapply to the same or to a different funder for a new grant. However, if implicit in the rejection of your grant proposal is the suggestion that other funders might not like it either, you may wish to turn to other means than grants to fund your idea.

LOANS

The possibility of seeking out a loan may not have occurred to you, or it may be unattractive to you because, eventually, a loan has to be paid back with interest. If, however, your idea is one that you expect will become profitable some day (e.g., you are an inventor), if it is one that will eventually be self-supporting (e.g., a service for which a given clientele will be willing to pay), if you only need start-up or seed money to get your project off the ground, or if your idea one day will be taken over by a larger institution (e.g., your sponsor or employer will be able to fund it next year), then you might consider applying for a loan.

Even in today's economy not all loans are at high interest rates. The federal government sometimes offers loans at a lesser or delayed interest rate for individuals in special categories. There is, for example, a program of low-interest Economic Opportunity Loans of $5,000 each for Vietnam veterans, and there are various programs for members of minority groups and women, especially in the area of new business enterprises. Government loans have the advantage of being guaranteed and direct.

Other resources for low-interest loans might be friends, relatives, wealthy individuals who take an interest in you or your idea, even your own employer. Commercial banks, of course, offer loans at prevailing market rates. Some are more willing to take risks than others. Another source for loans are small-business investment companies. If your idea qualifies, these companies have more flexible lending policies than commercial lenders.*

---

*If your idea qualifies as a small business and you want to seek a loan, see the very comprehensive book by Leonard E. Smollen, Mark Rollinson, and Stanley M. Rubel, *Source-guide for Borrowing Capital* (Wellesley Hills, Mass.: Capital Publishing Corp., 1977) and Richard C. Belew's *How to Negotiate a Business Loan* (New York: Van Nostrand Reinhold, 1973).

Occasionally you will hear about the availability of money to borrow through the grapevine in your local community or where you work. I am not speaking of loan sharks here, but of wealthy (often self-made) businessmen who are willing to take a chance by making a loan under a flexible pay-back arrangement to an individual whose idea strikes their fancy. To these potential lenders, the amount of the loan is relatively insignificant. (They might lose a similar amount on the stock market or at the race track in a single afternoon.) For these men, making such loans is their own form of philanthropy, although if you asked them, they would say they don't believe in charity, only in sound business investments.

Applying for a loan from a bank or other lender is an excellent exercise as either a precursor or a corollary to seeking a grant. It provides valuable insight for you into your own dollar needs and the fiscal requirements of your project. It also helps to build a good credit history and adds credibility for you as a responsible manager of your own financial affairs.

INVESTMENTS

There are instances when you may be able to get a corporation, your own employer or sponsor, or even a local businessman to treat your idea as an investment rather than as a grant or a loan. In such cases, you must be a pretty good salesman in order to interest potential investors in your idea.

Contrary to what you might expect, your idea need not be potentially profitable to have appeal as an investment. Some investors may be interested in your idea for tax reasons (e.g., they may need a financial loss for a given year). Others may be attracted to you or to your idea for more altruistic reasons, similar to those of the businessman mentioned in the last section who makes loans to worthy causes but simply refuses to give money away.

Some of the large private foundations have "program related investments" (PRI).[3] Under this type of program, a small percentage of the foundation's capital is available for investment for social purposes in projects related to the areas in which the foundation normally makes grants but that do not qualify for grant money (i.e., small businesses started by minorities or women, hospitals, clinics, consultation services that are not nonprofit, local action groups that may be controversial or political in nature). PRIs can take the form of no or low interest loans, guarantees, or purchases of preferred or other stock. They enable the foundation to support those recipients engaged in socially important activities who are ineligible for a grant because they are not tax-exempt. If you or your sponsor qualify as a PRI recipient, but you lack management expertise, the foundation's staff will provide you with technical advice and with help from accountants, lawyers, developers, and other professionals.

These are just a few examples of the ways in which your idea may be treated as an investment if you can't get a grant for it. As with loans, there is always some form of repayment implicit in the agreement to invest. To ascertain what this may entail, try to determine in advance the investor's true motivation and seek legal counsel, if necessary.

SELF-SUPPORT

If you've been turned down for a grant, or if your grant money has run out, you probably should step back, reexamine your idea, and see if there is some way to get it to support itself. Examples of means of self-support for your idea include product sales, publication royalties, subscriptions, membership fees, direct mail solicitation, fund-raising events, workshops, demonstration and training fees, and benefit or performance ticket sales, among others. Finding a way for your idea to support itself may require you to change and restructure it. However, such adaptations will prove worthwhile in the long run because you will end up answerable to no one but yourself.

Self-support is all a question of how ingenious you can be and how far you are willing to go. If your idea does not prove profitable, there are various ways that you can support yourself, short of full-time employment, while you continue working on your idea. These include entering your product or finished work in contests, festivals, competitions, salon exhibits, trade fairs, and other sales events. In addition to outright grants, there are internships, residencies, apprenticeships, trainee programs, artists' colonies, honoraria for speaking engagements, tutoring fees, and other personal benefit programs that enable an individual to support himself, including the bartering of products or services. You won't get rich on any of these, but they will help you to survive *if* working on your idea is important enough to you.

Other alternatives to getting grants are saving, investing, and moonlighting. However, all of these take time.

The concept behind finding alternative sources of funding is one of self-reliance. There is a vast self-help movement building in this country, based on the notion that self-reliance should replace government and other dispensers of money. To quote one proponent of the self-reliance movement: "If you are devoted enough and can find enough passion within yourself, you will find an almost infinite number of ways to make a living at the things you want to do."[4(p.10)]

NONFINANCIAL SUPPORT

Along with your search for ways in which your idea might support itself, you should consider ways that others can help you, besides giving you money. Many needs can be met without dollars, and you should learn to regard money as "only one of many resources."[5] Earlier in this chapter I spoke of viewing funders as resources for a wide range of services, not just suppliers of cash. The philosophy behind this concept is to turn a negative element (rejection of your grant proposal) into a positive one. If funders cannot give you outright grants, find a way that they *can* support you, be it via the provision of materials, tools, or simple advice. The solicitation of nonfinancial support from funders builds a relationship for the future since funding executives will become more interested and involved in your idea if they feel a part of it.

To identify other means of nonfinancial support, go back and reexamine the budget for your grant proposal. Are there services, materials, and equipment that, if donated by an outside source, will cut back on the cost of accomplishing your idea for a grant? Is it possible to get some corporation to release the time of an employee or a local organization to donate some volunteer time to perform certain tasks for which otherwise you would have to pay a salary or consultant fee (to yourself or someone else)?

Corporations, in particular, like to provide this type of nonfinancial support which they treat either as a business expense or as a gift in kind (see Chapter 5). Such activities might include commission of your work if you're an artist, purchase of finished products if you're a craftsman or inventor, free use of corporate facilities for exhibitions, performances, lectures, and access to computer, publishing, and research services for writers, scientists, and mathematicians.

As you begin to reconsider your proposal in light of soliciting various types of nonfinancial support, you may discover that it is possible to achieve part or even most of your idea without receiving a single grant dollar. As a consequence, you might find yourself with a brand new proposal on your hands that is smaller in scope and hence may be more attractive to funders. Furthermore, finding ways for others to help you out, even though it doesn't bring in dollars, will boost your own sagging morale and may give your grant idea new momentum.

BECOME A CONSULTANT

In Chapter 2 I spoke of becoming the employee of a nonprofit organization as an alternative to formal affiliation with a sponsor in order to seek grant money. Becoming a consultant is another side of the same coin and represents one alternative to seeking a grant. Consultants differ from employees in that they are paid fees for specific services provided by the hour, or day, rather than salaries based on the long-term provision of more general services. Consultants are far more independent than employees, and their advice is usually rather highly valued simply because they are "paid consultants."

Who can become a consultant? Almost anyone who has expertise in a given field or in the provision of a particular service can call himself a consultant. Sometimes becoming a consultant involves performing a service for a client; other times it involves teaching a client to do it for himself; still other times it means reviewing and making recommendations on activities the client presently conducts or on products or services the client is considering for the future.

To become a consultant you have to find someone (a local nonprofit organization, corporation, educational institution, even a government funding agency) to pay you a fee to do what you originally wanted a grant to do. To launch you on your new career, you require only one satisified client, who, in turn, may recommend you to others or whose name you can use to help get other clients.

What does all of this have to do with getting a grant? In some cases the difference between a grant and a fee might be purely one of semantics. In other instances, the same funder that turned down your grant idea might be willing to hire you on a consultant basis to conduct the same or a similar project, *if* the idea is perceived as important by funding management. The concept behind such transactions, of course, is that grant money is frequently more tied up and more closely scrutinized than administrative funds.

Examples of consulting services that individuals may provide for a fee are poets and writers conducting workshops, giving lectures, and readings, artists designing posters and assisting with advertising campaigns, researchers conducting experiments, technicians providing training, musicians and other performers supplying entertainment. Teachers and specialists in all kinds of areas might become consultants and receive a fee for accomplishing their ideas if they can convince the client that the services they provide are necessary to the achievement of its institutional goals.

Finding someone to pay you for doing what you really want to do represents almost everyone's ideal. It is truly the determining factor for success in grant seeking as well as in many other professional endeavors, and it may require great resourcefulness on your part.

By now it probably has become apparent that if you remain flexible in your consideration of various alternatives to seeking a grant, the distinctions among the roles of funder, sponsor, employer, and client become less clearcut. In fact, if you are ingenious, you may find a way for a given organization to play any one of these roles or several of them sequentially or simultaneously, *if* you can get their leadership interested enough in you and your idea. At the same time, the differences among grants and loans, salaries, commissions, and consulting fees lose some of their importance to you. It may be that it really doesn't matter *how* you get the money, as long as it is legal, of course, *and* it enables you to accomplish what you set out to do. The final decision rests with you about how far you are willing to go, and how many compromises you are willing to make along the way in order to fund your idea.

WHAT ELSE CAN BE GAINED FROM THE GRANT-SEEKING PROCESS?

Earlier in this chapter I mentioned that you can get more from funders than just grants. Similarly, there is much you can derive from grant seeking besides money. You might consider many of these benefits as "consolation" or "booby prizes" for the loser. Surely, this is too negative a view. If you have followed the grant-seeking process through from start to finish, whether or not you are funded, you will have discovered numerous opportunities for self-examination, for redefining your goals and for developing your creative abilities and ideas. For its growth potential alone, grant seeking is well worth the effort.

In an early memorandum on grants to individuals, an executive of Carnegie Corporation stated that at the end of his grant-seeking endeavor the individual should find himself "neither richer nor poorer."[6] I disagree. At the end of this process, whether or not you actually are funded, you should find yourself far richer in terms of the learning experience and the contacts you have made.

Even if a particular proposal is unsuccessful, when you present your idea to potential funders, you derive the advantage of increased visibility, and you learn more about funders' priorities and activities. You secure advice about strengthening your proposal in order to find alternate sources of funding and to explore ways to cooperate with funders on future ventures. Whether or not you persist in your attempt to fund your present idea, revise it, or go on to an entirely different one, you will find yourself much better prepared now that you once have gone through the process.

As a second consolation, consider the proposition that creativity in and of itself may be its own reward. To quote Rollo May: "If one is preoccupied with thoughts of applause, approval, or reward, the process (of creativity) is short circuited." May also states: "Giving is the essence of creativity."[7] For the grant seeker, his recommendation might be to view your idea for a grant as a generous act, a gift that you proffer to your community, whether or not you are reimbursed for it.

Now you might be tempted to protest: But if you can't get the money to execute your idea, what good is it anyway? The answer is to keep in mind that your idea has intrinsic value of its own. If you truly believe in it, then eventually you may find a way to carry it out. In the opinion of at least one successful inventor, the very worst thing is to leave your idea untested because you were afraid to take a chance on it.[8]

This reminds you once again that grant seeking is a risky business— it leaves you exposed to criticism and, even worse, rejection. However, learning to take criticism and turning it to your own advantage is a quality to be cultivated. It can lead to success in almost any endeavor you undertake. To quote an engineer/psychologist who specializes in problem-solving techniques: "The self actualizer, the expert, the genius—all capitalize on their mistakes. Errors are pretty much a part of growth."[9]

As one final consolation, you might consider the revolutionary proposition that money may not be all that important. It may be viewed as secondary to whatever project you are working on at a given time. Money is neither a god nor a demon but just another tool. All too often, the grant seeker permits money (or its absence) to serve as an obstacle preventing him from doing what he really wants to do. Perhaps doing something well (or just part of your project well) is enough, even if you don't get paid for it.

In his book, *The Seven Laws of Money,* Michael Phillips (President of the Glide Foundation and originator of *The Whole Earth Catalog*) states as his first law: "Do It! Money will come when you are doing the right thing."[4(p.1)] Although this assertion may seem overly naive and optimistic to anyone who has tried seeking a grant, it does bring out an important point: Money serves as a major stumbling block to the carrying out of an idea. It is wise, then, to put money back in perspective and not let the lack of it or the search for it get in the way of your real work. The moral is not to be discouraged if your first attempt at getting a grant ends in failure. If you can accept Phillips' first law, you can stop worrying about seeking a grant and get on with your idea. Adopt the credo that if it's worth doing, you'll probably find a way to do it. According to one of Phillips' cronies: "People energy is far more important than any grant and a good project will always be provided for."[10]

LUCK

In the chapter on getting good ideas are several examples of fortunate individuals who simply were in the right place at the right time. Coincidence, timing, and adopting a receptive attitude are all elements of having good luck. Of course, genius doesn't hurt either. But in the end, achieving success in grant seeking (and in life) may be mostly a matter of fate rather than talent![10] After all, you may be brilliant and have an innovative idea, but those you present it to must be receptive in order for you to attain success.

Luck pure and simple does play a role in determining whether or not you receive a grant. Sometimes a funder may be searching for new ideas, initiating new programs, or even disbursing surplus funds at just the moment you decide to submit your grant proposal. Or if you are fortunate enough to be called for an interview, you may discover that you come from the same hometown as an important grants decision maker. You and the interviewer may just "click" in some other way. Subjective elements like chemistry cannot help but influence the grants decision-making process.

Is there anything you can do to improve your luck? Yes, keep on trying. Seek out funding opportunities, develop new ideas, and formulate proposals. Continue to submit them to potential funders. An idea that never finds its way onto a piece of paper or a proposal that sits in your desk drawer will not be funded, short of a miracle. However, if you persist in your determination to receive a grant for your idea, if you continue to put your ideas forward for funders' consideration, eventually you should get lucky. At least you will have increased the odds in your favor.

CONCLUSION: COMMUNICATION IS THE WAY

I have dwelt on various reactions to the rejection of your proposal because, almost inevitably, at some point grant seekers will come up against rejection. It is possible to view the rejection of a grant request simply as failure to communicate both on the part of the grant seeker and the funder. I would like to end this book with a fervent plea for increased communication, addressed to both groups.

To the individual grant seeker: How will funders know your needs and desires unless you ask questions and make your requests known? To the funding executive: How will individual grant applicants know what types of projects you fund and what you expect in a proposal unless you tell them? In the future, funders should consider the publication of uniform detailed guidelines for the benefit of individual applicants. To quote one grants officer whose foundation makes grants to individuals: "The obligation to inform is a shared burden."[11]

Grant seekers and funding executives traditionally have tended to view their relationship as adversarial. In my opinion, this is one reason for the high turndown rate of grant proposals. Both grant seekers and funding executives may be victims of stereotypical thinking and erroneous attitudes about one another. They share a mutual misunderstanding and mistrust. Individuals tend to view funders as monolithic soulless enterprises dedicated to maintaining the status quo. They perceive the operating objectives of funding institutions as vague and subject to reversal at the donor's whim. They ask, how can you play the game when you don't know the rules? Many funding executives view individual applicants as eccentric bohemians, unreliable and incapable of administering programs involving substantial sums of money because they are untrained and disdainful of financial matters. Funders must continually prod themselves so they won't be inhospitable to people who wish to follow lonely trails (e.g., propose unfashionable ideas). However, first they must concede that this tendency exists and then give more to stimulate individual work.[12]

These are the all-too-familiar stereotypes of the clash between the artist and the businessman. The way to combat these stereotypes is to increase and improve the transfer of accurate information about one another to one another. In his book, *The Writer on His Own,* David Greenhood states: "When the critical and the creative are not afraid of each other and not overworshipful, they get along together as united forces."[13] Substitute funder for "critical" and grant seeker for "creative" and you have a truism about the grant-seeking process as well.

Information and communication between those on both sides of the grant dollar are the vehicles to the improved attitudes necessary to bring grant seekers and funders closer together. Fear of the unknown and reluctance to take risks are most destructive to the grantsmanship process. It is my hope that this book represents one step in the direction of tearing down the walls between the two groups. If the seekers and the givers begin to communicate more openly and effectively with one another, everyone will benefit. The quality of applicants will improve. Better, more sophisticated and thoughtful proposals will be submitted to more appropriate, hence more responsive, funders. Eventually, this will result in more successful, diverse, and potentially more innovative grants being awarded to increasing numbers of individual applicants.

NOTES

[1] Smith, C. W., & Skjei, E. W. *Getting grants.* New York: Harper & Row, 1980. P. 162.

[2] Brownrigg, W. G. *Corporate fund raising.* New York: American Council for the Arts, 1978. P. 42.

[3] *Program-related investments: A different approach to philanthropy.* New York: Ford Foundation, 1974.

[4] Phillips, M. *The seven laws of money.* Menlo Park, Calif./New York: Word Wheel/Random House, 1974.

[5] Mitiguy, N. *The rich get richer and the poor write proposals.* Amherst: Univ. of Massachusetts, Citizens Involvement Training Project, 1978. P. 141.

[6] Dollard, C. *Memorandum on grants to individuals.* New York: Carnegie Corp., 1938. P. 13.

[7] Rollo May in Rockmere, M. Creating creativity, *American Way,* March 1978, p. 87.

[8] Kracke, D., with Honkanen, R. *How to turn your idea into a million dollars.* New York: Doubleday, 1977.

[9] Melvin, T. *Practical psychology in construction management.* New York: Van Nostrand Reinhold, 1979. P. 279.

[10] Rasberry, S. Introduction. In Phillips,[4] p. v.

[11] Hartwig, J. C. Betting on foundations. In *Foundation grants to individuals* (2nd ed.). New York: The Foundation Center, 1979. P. xiii.

[12] Whyte, W. H., Jr. *The organization man.* New York: Simon & Schuster, 1956. P. 238.

[13] Greenhood, D. *The writer on his own.* Albuquerque: Univ. of New Mexico Press, 1977. P. 177.

Appendix / FEDERAL
INFORMATION
CENTERS

State/city	Address	Telephone/Tieline
Alabama		
Birmingham		322-8591
Mobile		438-1421
Alaska		
Anchorage	Federal Bldg. and U.S. Courthouse, 701 C St., 99513	907-271-3650
Arizona		
Tucson		622-1511
Phoenix	Federal Bldg., 230 N. 1st Ave., 85025	602-261-3313
Arkansas		
Little Rock		378-6177
California		
Los Angeles	Federal Bldg., 300 N. Los Angeles St., 90012	213-688-3800
Sacramento	Federal Bldg., U.S. Courthouse, 650 Capitol Mall, 95814	916-440-3344
San Diego	Government Information Center, Federal Bldg., 880 Front St., 92188	714-293-6030
San Francisco	Federal Bldg., U.S. Courthouse, 450 Golden Gate Ave., 94102	415-556-6600

San Jose		275-7422
Santa Ana		836-2386
Colorado		
Colorado Springs		471-9491
Denver	Federal Bldg., 1961 Stout St., 80294	303-837-3602
Pueblo		544-9523
Connecticut		
Hartford		527-2617
New Haven		624-4720
Florida		
Fort Lauderdale		522-8531
Jacksonville		354-4756
Miami	Federal Bldg., 51 S.W. 1st Ave., 33130	305-350-4155
Orlando		422-1800
St. Petersburg	Cramer Federal Bldg., 144 1st Ave. S., 33701	813-893-3485
Tampa		229-7911
West Palm Beach		833-7566
Other south Florida locations		800-432-6668

[a]The tieline numbers are toll-free, only in the specified city.

State/city	Address	Telephone/Tieline
Other north Florida locations		800-282-8556
Georgia Atlanta	Russell Federal Bldg., U.S. Courthouse, 75 Spring St. S.W., 30303	404-221-6891
Hawaii Honolulu	300 Ala Moana Blvd., 96850	808-546-8620
Illinois Chicago	Everett McKinley Dirksen Bldg., Room 250, 219 S. Dearborn St., 60604	312-353-4242
Indiana Gary/Hammond Indianapolis	Federal Bldg., 575 N. Pennsylvania St., 46204	883-4110 317-269-7373
Iowa Des Moines Other Iowa locations	Federal Bldg., 210 Walnut St., 50309	515-284-4448 800-532-1556

Kansas		
Topeka	Federal Bldg., U.S. Courthouse, 444 S.E. Quincy, 66683	913-295-2866
Other Kansas locations		800-432-2934
Kentucky		
Louisville	Federal Bldg., 600 Federal Pl., 40202	502-582-6261
Louisiana		
New Orleans	U.S. Custom House, Room 100, 423 Canal St., 70130	504-589-6696
Maryland		
Baltimore	Federal Bldg., 31 Hopkins Plaza, 21201	301-962-4980
Massachusetts		
Boston	John F. Kennedy Federal Bldg., Room E-121, Cambridge St., 02203	617-223-7121
Michigan		
Detroit	McNamara Federal Bldg., 477 Michigan Ave., 48226	313-226-7016
Grand Rapids		451-2628

State/city	Address	Telephone/Tieline
Minnesota		
Minneapolis	Federal Bldg.–U.S. Courthouse, 110 S. 4th St., 55401	612-349-5333
Missouri		
Kansas City	Federal Bldg., 601 E. 12th St., 64106	816-374-2466
St. Louis	Federal Bldg., 1520 Market St., 63103	314-425-4106
Other Missouri locations within area code 314		800-392-7711
Other Missouri locations within area codes 816 and 417		800-892-5808
Nebraska		
Omaha	U.S. Post Office and Courthouse, 215 N. 17th St., 68102	402-221-3353
Other Nebraska locations		800-642-8383
New Jersey		
Newark	Federal Bldg., 970 Broad St., 07102	201-645-3600

Paterson/Passaic		523-0717
Trenton		396-4400
New Mexico		
Albuquerque	Federal Bldg.–U.S. Courthouse, 500 Gold Ave. S.W., 87102	505-766-3091
Santa Fe		983-7743
New York		
Albany		463-4421
Buffalo	Federal Bldg., 111 W. Huron St., 14202	716-846-4010
New York	Federal Office Bldg., 26 Federal Plaza, 10007	212-264-4464
Rochester		546-5075
Syracuse		476-8545
North Carolina		
Charlotte		376-3600
Ohio		
Akron		375-5638
Cincinnati	Federal Bldg., 550 Main St., 45202	513-684-2801
Cleveland	Federal Bldg., 1240 E. 9th St., 44199	216-522-4040
Columbus		221-1014
Dayton		223-7377
Toledo		241-3223

State/city	Address	Telephone/Tieline
Oklahoma		
Oklahoma City	U.S. Post Office and Courthouse, 201 N.W. 3rd St., 73102	405-231-4868
Tulsa		584-4193
Oregon		
Portland	Federal Bldg., Room 109, 1220 S.W. 3rd Ave., 97204	503-221-2222
Pennsylvania		
Philadelphia	William J. Green, Jr., Federal Bldg., 600 Arch St., 19106	215-597-7042
Allentown/Bethlehem		821-7785
Pittsburgh	Federal Bldg., 1000 Liberty Ave., 15222	412-644-3456
Scranton		346-7081
Rhode Island		
Providence		331-5565
Tennessee		
Chattanooga		265-8231
Memphis	Clifford Davis Federal Bldg., 167 N. Main St., 38103	901-521-3285

Nashville		242-5056
Texas		
Austin		472-5494
Dallas		767-8585
Fort Worth	Fritz Garland Lanham Federal Bldg., 819 Taylor St., 76102	817-334-3624
Houston	Federal Bldg.–U.S. Courthouse, 515 Rusk Ave., 77002	713-226-5711
San Antonio		224-4471
Utah		
Ogden		399-1347
Salt Lake City	Federal Bldg., Room 1205, 125 S. State St., 84138	801-524-5353
Virginia		
Newport News		244-0480
Norfolk	Federal Bldg., Room 120, 200 Granby Mall, 23510	804-441-3101
Richmond		643-4928
Roanoke		982-8591
Washington		
Seattle	Federal Bldg., 915 2nd Ave., 98174	206-442-0570
Tacoma		383-5230
Wisconsin		
Milwaukee		271-2273

FURTHER READING

CHAPTER 1: INTRODUCTION

Andrews, F. E. *Philanthropic giving.* New York: Russell Sage Foundation, 1950.

Annual register of grant support. Chicago: Marquis Academic Media (See introductory materials to latest edition).

Dollard, C. *Memorandum on grants to individuals.* New York: Carnegie Corp., 1938.

Dressner, H. R. Don't stop making grants to individuals. *Foundation News,* September/October 1973, pp. 26–30.

Hartwig, J. C. Betting on foundations. In *Foundation grants to individuals* (2nd ed.). New York: The Foundation Center, 1979. Pp. xii–xiv.

CHAPTER 2: NO MAN IS AN ISLAND

Burns, J. S. *The awkward embrace, The creative artist and the institution in America.* New York: Alfred A. Knopf, 1975.

Greenhood, D. *The writer on his own.* Albuquerque: Univ. of New Mexico Press, 1971.

McCarthy, J. T., Bauer, J. C., & Call, R. I. Successful grantsmanship. *Health Care Management Review,* Summer 1978, pp. 37–44.

McLuhan, M., & Fiore, Q. *The medium is the massage.* New York: Random House, 1967.

Montagu, A. The crisis in scientific research. *Foundation News,* July/August 1980, pp. 18–19.

Non-profit cultural organizations, Patents, Copyrights, Trademarks, and Literary Property Course Handbook Series, No. 113. New York: Practicing Law Institute, 1979.

Presthus, R. *The organizational society.* New York: St. Martin's Press, 1978.

Reisman, D. *The lonely crowd, A study of the changing American character.* New Haven, Conn.: Yale Univ. Press, 1950.

Reisman, D. *Individualism reconsidered and other essays.* Glencoe, Ill.: The Free Press, 1954. Chap. 18, pp. 266–272.

Smith, C. W., & Skjei, E. W. *Getting grants.* New York: Harper & Row, 1980. Chap. 4, pp. 147–168.

Sponsored research policy of colleges and universities, A report of the Committee on Institutional Research Policy. Washington, D.C.: American Council on Education, 1954.

U.S. Department of the Treasury. *How to apply for and retain exempt status for your organization* (IRS Publication 557). Washington, D.C.: U.S. Govt. Printing Office, 1981.

Weaver, P. *You, Inc., A detailed escape route to being your own boss.* New York: Doubleday, 1973.

Whitaker, F. A. *How to form your own non-profit corporation in one day.* Oakland, Calif.: Minority Management Institute, 1978.

Whyte, W. H., Jr. *The organization man.* New York: Simon & Schuster, 1956. Chap. 8, pp. 230–240.

Wolf, T. *Presenting performances, A handbook for sponsors* (2nd ed.). Cambridge, Mass.: New England Foundation for the Arts, 1977.

CHAPTER 3: IN THE BEGINNING IS AN IDEA

Adams, J. L. *Conceptual blockbusting, A guide to better ideas.* San Francisco: Freeman, 1974.

Barron, F. *Creativity and psychological health, Origins of personal vitality and creative freedom.* New York: D. Van Nostrand, 1963.

Belcher, J. C., & Jacobsen, J. M. *A process for development of ideas.* Washington, D.C.: News, Notes, and Deadlines, 1976.

Cooper, M. *The inventions of Leonardo Da Vinci.* New York: Macmillan, 1965.

de Bono, E. *New think, The use of lateral thinking in the generation of new ideas.* New York: Basic Books, 1968.

de Bono, E. *Lateral thinking, Creativity step by step.* New York: Harper Colophon Books, 1970.

Glassman, D. *Writers' and artists' rights, Basic benefits and protections to authors, artists, composers, sculptors, choreographers and movie-makers under the new American copyright law.* Washington, D.C.: Writers Press, 1978.

Groh, J. *The right to create.* Boston, Little, Brown, 1969.

Grossinger, T. *The book of gadgets.* New York: McKay, 1974.

Guth, H. P. *Words and ideas, A handbook for college writing* (4th ed.). Belmont, Calif.: Wadsworth, 1975. Chap. 7, pp. 150–177.

Heyn, E. V. *Fire of genius, Inventors of the past century.* New York: Doubleday, 1976.

Hooper, M. *Everyday inventions.* London: Augus & Robertson, 1972.

Kivenson, G. *The art and science of inventing.* New York: Van Nostrand Reinhold, 1977.

Koestler, A. *The art of creation.* New York: Macmillan, 1964.

Kracke, D., with Honkanen, R. *How to turn your idea into a million dollars.* New York: Doubleday, 1977.

May, R. *The courage to create.* New York: Norton, 1975.

McNair, E. P. and Schwenck, J. E. *How to become a successful inventor.* New York: Hastings House, 1973.

Melvin, T. *Practical psychology in construction management.* New York: Van Nostrand Reinhold, 1979. Chap. 7, pp. 273–326.

Moohey, R. L., & Razik, T. A. (Eds.). *Explorations in creativity.* New York: Harper & Row, 1967.

Pauling, L. Reforming the grant system. *Foundation News,* July/August 1980, pp. 19–22.

Rockmore, M. Creating creativity. *American Way,* March 1979, pp. 83–87.

Yalow, R. The dilemma of funding. *Foundation News,* July/August 1980, pp. 24, 43.

CHAPTER 4: GETTING INTO GEAR

American Association of University Women. *Community action tool catalog, Techniques and strategies for successful action programs.* Washington, D.C.:1978.

Fromm, E. *Escape from freedom.* New York: Holt, Rinehart & Winston, 1941.

Girard, J., with Casemore, R. *How to sell yourself.* New York: Simon & Schuster, 1979.

Knaus, W. J. *Do it now, How to stop procrastinating.* Englewood Cliffs, N.J.: Prentice-Hall, 1979.

Michaels, D. M., & Magrand, L. E., Jr. *A guide to grantsmanship.* Durham, N. H.: New England Gerontology Center, 1977.

Winston, S. *Getting organized, The easy way to put your life in order.* New York: Norton, 1978.

CHAPTER 5: FACTS ABOUT FUNDERS

GENERAL

Andrews, F. E. *Philanthropy in the United States, History and structure.* New York: The Foundation Center, 1978.

Bellows, B. The grants game. *Writers' Digest,* January 1981, pp. 35–37.

Commission on Private Philanthropy and Public Needs. *Giving in America, Toward a stronger voluntary sector.* Washington, D.C.: 1975.

Giving U.S.A.—1982 Annual Report, A compilation of facts and trends on American philanthropy for the year 1981. New York: American Association of Fund Raising Counsel, 1982.

Jenkins, E. C. *Philanthropy in America, An introduction to the practices and prospects of organizations supported by gifts and endowments, 1924–1948.* New York: Association Press, 1950.

Lee, L. *The grants game, How to get free money.* San Francisco: Harbor Publishing, 1981.

Schneider, P. H. *The art of asking, Handbook for successful fund raising.* New York: Walker, 1978.

White, V. *Grants, How to find out about them and what to do next.* New York: Plenum, 1975.

GOVERNMENT

Hillman, H. *The art of winning government grants.* New York: Vanguard Press, 1977.

Holtz, H. *Government contracts, Proposalmanship and winning strategies.* New York: Plenum, 1979.

Johnson, W., & Colligan, F. J. *The Fulbright program, A history.* Chicago: Univ. of Chicago Press, 1965.

National Endowment for the Humanities. *Fifteenth annual report 1980.* Washington, D.C.: 1981.

National Health Council. *An introduction to grants and contracts in major HEW health agencies* (2nd ed.). New York: 1978.

National Science Foundation. *Grants for scientific research* (Rev. ed.). Washington, D.C.: U.S. Govt. Printing Office, 1980.

Nielsen, W. A. Needy arts: Where have all the patrons gone? *New York Times,* October 26, 1980, Section 2, pp. 1; 26–27.

Operation revenue, Orientation resource notebook for governmental activities. Washington, D.C.: United Cerebral Palsy Association, Inc., 1976.

Orr, S. Coping with block grants. *Grantsmanship Center News,* November/ December 1981, pp. 40–43.

Safire, W. The newest federalism. *New York Times,* January 28, 1982, p. A23 (Op Ed).

Straight, M. *Twigs for an eagle's nest, Government and the arts 1965–1978.* New York: Devon Press, 1979.

FOUNDATIONS

Andrews, F. E. *Philanthropic foundations.* New York: Russell Sage Foundation, 1956.

The foundation directory (8th ed.). New York: The Foundation Center, 1981. Introductory materials.

Foundation grants to individuals (3rd ed.). New York: The Foundation Center, 1982.

Foundations, private giving and public policy, Report and recommendation of the Commission on Foundations and Private Philanthropy (Peterson Commission). Chicago: Univ. of Chicago Press, 1970.

Thomson, E. G. Federal law and grants to individuals. In *Foundation grants to individuals* (3rd ed.). New York: The Foundation Center, 1982. Pp. xiii–xiv.

Zurcher, A. J., & Dustan, J. *The foundation administrator, A study of those who manage America's foundations.* New York: Russell Sage Foundation, 1972.

CORPORATIONS

Aid to education programs of leading business concerns. New York: Council for Financial Aid to Education, 1978.

Boyd, T. The business of business is . . . philanthropy? *Grantsmanship Center News,* May/June 1982, pp. 56–63.

Brownrigg, W. G. *Corporate fund raising, A practical plan of action.* New York: American Council for the Arts, 1978.

Carr, E. G., & Morgan, J. F. *Better management of business giving.* New York: Hobbs, Dorman, 1966.

Chagy, G. (Ed.). *Business in the arts '70.* New York: Eriksson, 1969.

Chagy, G. (Ed.). *The state of the arts and corporate support.* New York: Eriksson, 1971.

Chamberlin, N. N. *The limits of corporate responsibility.* New York: Basic Books, 1973.

Fremont-Smith, M. R. *Philanthropy and the business corporation.* New York: Russell Sage Foundation, 1972.

Hillman, H. *The art of winning corporate grants.* New York: Vanguard Press, 1980.

Jacoby, N. H. *Corporate power and social responsibility, A blueprint for the future.* New York: Macmillan, 1973.

Mahne, H. G., & Wallich, H. C. *The modern corporation and social responsibility.* Washington, D.C.: American Enterprise Institute for Public Policy Research, 1972.

McKie, J. W. (Ed.). *Social responsibility and the business predicament.* Washington, D.C.: The Brookings Institution, 1974.

Murphy, D. J. *Asking corporations for money.* Port Chester, N.Y.: Gothic Press, 1982.

1978 Social report of the life and health insurance business. Washington, D.C.: Clearinghouse on Corporate Social Responsibility, 1978.

Shakeley, J. Exploring the elusive world of corporate giving. *Grantsmanship Center News,* July/September 1977, pp. 35–59.

Simon, J. G., Powers, C. W., & Gunnemann, J. P. *The ethical investor.* New Haven, Conn.: Yale Univ. Press, 1972.

Troy, K. *Annual survey of corporate contributions* (Report No. 802). New York: The Conference Board, 1981.

Wagner, S. E. *A guide to corporate giving in the arts.* New York: American Council for the Arts, 1978.

CHAPTER 6: FINDING THE RIGHT FUNDER FOR YOU

See also the two-part "Annotated Bibliography of Printed Sources on Grants to Individuals," included in Chapter 6.

Barzun, J., & Graff, H.F. *The modern researcher,* (Rev. ed.). New York: Harcourt, Brace, 1957.

Broce, T. E. *Fund raising, The guide to raising money from private sources.* Norman: Univ. of Oklahoma Press, 1979.

Fund Raising: A guide for research on individuals. Los Angeles: Gary W. Phillips and Associates, 1978.

Glass, S. A. A winning strategy; for victory in the home stretch, spend more time cultivating foundations and less mailing proposals." *Case Currents,* May 1980, pp. 29–31.

Heaton, W. E. Research: Key to fund raising success. *Philanthropy Monthly,* June 1980, pp. 22–25.

Kohl, K. A. & Kohl, I. C. *Financing college education.* New York: Harper & Row, 1980.

Koochoo, E. *Prospecting, Searching out the philanthropic dollar.* Washington, D.C.: Taft Corp. 1979.

Kurzig, C. M. *Foundation fundamentals.* New York: The Foundation Center, 1980.

Law-Yone, W. *Company information, A model investigation.* Washington, D.C.: Washington Researchers, 1980.

Margolin, J. B. *About foundations, How to find the facts you need to get a grant* (Rev. ed.). New York: The Foundation Center, 1977.

Zallen, H. & Zallen E. M. *Ideas plus dollars, Research methodology and funding.* Norman, Okla.: Academic World, 1976.

CHAPTER 7: THE PROPOSAL

Conducting evaluations, Three perspectives. New York: The Foundation Center, 1980.

The Conference Board. *Evaluating new product proposals.* New York: 1973.

Conrad, D. L. *The grants planner.* San Francisco: The Institute for Fund Raising, 1976.

Dermer, J. *How to write successful foundation presentations.* New York: Public Service Materials Center, 1975.

Hall, M. *Developing skills in proposal writing* (2nd ed.). Portland, Ore.: Continuing Education Publications, 1977.

Kiritz, N. J. *Program planning and proposal writing* (expanded version). Los Angeles: The Grantsmanship Center, 1979. Pp. 33–79 (reprint).

Lefferts, R. *Getting a grant, How to write successful grant proposals.* Englewood Cliffs, N.J.: Prentice–Hall, 1978.

Warm, H. How grant-seekers lower the odds. *Foundation News,* July/August 1979, pp. 21, 37–39, 43.

CHAPTER 8: FOLLOW-UP

Arkin, J. *Taxation of prizes, awards and scholarships.* New York: Reiner Press, 1962.

Bolles, R. N. *What color is your parachute? A practical manual for job hunters and career changers* (Rev. ed.). Berkeley, Calif.: Ten Speed Press, 1982.

Corey, K. E. *Neighborhood grantsmanship, An approach for grassroots self-reliance in the 1980's.* Cincinnati, Ohio: Community Human Resources Training, Inc., 1979.

Fear of filing. New York: Volunteer Lawyers for the Arts, 1981.

Holtzheiman R. *Profit from your money-making ideas, How to build a new business or expand an existing one.* New York: AMACOM, 1980.

Levin, N. J., & Steiger, J. D. *To light one candle, A handbook for organizing, funding and maintaining public service activities.* Washington, D.C.: American Bar Association, 1978.

Marcuso, J. R. *How to start, finance, and manage your own small business.* Englewood Cliffs, N.J.: Prentice–Hall, 1978.

Meredith, S. *Writing to sell.* New York: Harper & Row, 1950.

Mitiguy, N. *The rich get richer and the poor write proposals.* Amherst: Univ. of Massachusetts Citizen Involvement Training Project, 1978.

Nicholas, T. *Where the money is and how to get it.* Wilmington, Dela.: Enterprise Publishing, 1980.

Phillips, M. *The seven laws of money.* Menlo Park, Calif./New York: World Wheel/Random House, 1974.

Program related investments, A different approach to philanthropy. New York: Ford Foundation, 1974.

Scholarships, grants, stipends and taxes, too. Chicago: Commerce Clearing House Inc., 1977.

Smollen, L. E., Rollinson, M., & Rubel, S. M. *Sourceguide for borrowing capital.* Wellesley Hills, Mass.: Capital, 1977.

INDEX